Asteroid Rendezvous
NEAR Shoemaker's Adventures at Eros

The Near Earth Asteroid Rendezvous (NEAR) was the first mission to orbit and eventually land on an asteroid. NEAR was also one of the first tests of NASA's new "better, faster, cheaper" planetary exploration philosophy. The mission was a phenomenal success, returning hundreds of thousands of images, spectra, and other measurements about the large near-Earth asteroid 433 Eros. This book is a collection of essays by some of the scientists and engineers who made NEAR such a success. The entire mission is described here in their own words, from the initial concept studies, through the development phase, launch, cruise operations, the flyby of asteroid Mathilde, the near-catastrophic main engine failure in 1998, the heroic rescue and recovery of the spacecraft, the amazing year-long up-close look at one of our most primitive celestial neighbors, and finally the daring attempt to land the spacecraft on Eros at the end of the mission. The book is liberally illustrated throughout with images from the mission and explanatory diagrams.

JIM BELL is a planetary scientist and professor in the Department of Astronomy at Cornell University. His research specialty is the composition and geology of planetary surfaces and he has participated in many NASA space science missions, including Mars Pathfinder and NEAR. Author of more than 70 first and co-authored journal publications and book chapters, he is also a frequent contributor to popular astronomy magazines and radio shows. The International Astronomical Union recently awarded him the honor of having asteroid 8146 Jimbell named after him.

JACQUELINE MITTON trained as an astronomer (MA, Oxford, physics; Ph.D., Cambridge, astrophysics). For about ten years she has been a full-time writer and media consultant specializing in astronomy. She has been the Press Officer of the Royal Astronomical Society since 1989, and was Editor of the *Journal of the British Astronomical Association*, 1989–1993. She is the author or co-author of 17 published books. Asteroid 4027 Mitton was named for Jacqueline and her husband, Simon.

Asteroid Rendezvous

NEAR Shoemaker's Adventures at Eros

Edited by

Jim Bell

and

Jacqueline Mitton

CAMBRIDGE
UNIVERSITY PRESS

PUBLISHED BY THE PRESS SYNDICATE OF THE UNIVERSITY OF CAMBRIDGE
The Pitt Building, Trumpington Street, Cambridge, United Kingdom

CAMBRIDGE UNIVERSITY PRESS
The Edinburgh Building, Cambridge CB2 2RU, UK
40 West 20th Street, New York, NY 10011–4211, USA
477 Williamstown Road, Port Melbourne, VIC 3027, Australia
Ruiz de Alarcón 13, 28014 Madrid, Spain
Dock House, The Waterfront, Cape Town 8001, South Africa

http://www.cambridge.org

© Cambridge University Press 2002

First published 2002

Printed in the United Kingdom at the University Press, Cambridge

Typeface Rotis   *System* QuarkXPress® [SE]

A catalogue record for this book is available from the British Library

ISBN 0 521 81360 3 hardback

Contents

Preface and acknowledgements

Space exploration is a risky affair. Spacecraft, and the sensitive scientific instruments they carry, have to survive violent shocks and vibrations during launch, the hostile cold and vacuum of the space environment, and long-duration exposure to high-energy solar wind particles, galactic cosmic rays, and/or intense planetary magnetic fields. Add in the hardware and software of specialized computers and the potential for human error in any complex project, and it is incredible that we have been able to successfully explore space with our robotic and human-piloted machines at all.

The NEAR mission was conceived and designed over the course of more than a decade, based on input from a number of NASA and National Academy of Sciences recommendations about the next steps required to fundamentally increase our knowledge about primitive solar system objects. The nuts and bolts and wires started being assembled in 1993, and the spacecraft was launched in early 1996. The nominal mission plan called for an Earth gravity-assist flyby in 1998, followed by a one-year rendezvous and orbital mission at the large near-Earth asteroid 433 Eros beginning in early 1999. Aside from the fairly routine launch and early cruise operations, however, the mission was anything but nominal. It was laced with surprises, ranging from the near-catastrophic failure of the spacecraft's main engine during a critical maneuver when first attempting to enter Eros orbit, to the unexpected and delightful survival of the spacecraft on the surface of Eros after it was landed on the asteroid at the end of the mission.

Redundancy, testing, proper contingency planning, and luck have often been cited as factors that offset the many potential pitfalls inherent in space exploration. As this book describes, each of these factors played a critical role in NASA's successful Near Earth Asteroid Rendezvous (NEAR) mission. Redundancy in the spacecraft design was exploited by using the second solid state data recorder on the spacecraft to double the number of images and other data that could be obtained. Testing was extensive during preflight and cruise activities, and helped identify a number of software and hardware glitches that could have been severely debilitating if not found until later in the mission. Contingency planning (no-one's favorite job: who likes to plan for failures?) proved absolutely critical when a major failure occurred, because most of the calculations for a second attempt at a main engine burn had already been performed. And, finally, many insiders involved with the mission believe that it was only by chance that the spacecraft corrected its wild spinning and gyrating and locked its antenna signal on Earth after the main engine misfire in 1998. Whatever the origin and combination of forces at work, one thing is certain: NEAR Shoemaker (as it was later named) proved to be a resilient and flexible spacecraft that ultimately responded well to the rigors of daily operations and the harsh environment of space.

The events that unfolded during NEAR Shoemaker's five-year interplanetary odyssey were a roller coaster of emotional and scientific ups and downs for the scientists and engineers directly involved in the mission. Here, in this book, we have attempted to capture some of that spirit in a collection of personal essays on various aspects of the mission, written by some of those directly involved. We take you, our readers, on a first-hand journey of exploration from the initial mission design and fabrication process, through launch and cruise operations, including a fortuitous flyby of the asteroid 253 Mathilde in 1997 and the Earth gravity assist in 1998, the failed initial orbital rendezvous attempt in late 1998, the wildly successful orbital operations phase throughout 2000, and finally the daring landing on the asteroid itself in early 2001.

The chapters are not formal papers describing the mission and its results in a systematic way but are rather a series of stories and vignettes about the inner workings of a complex space mission, told by some of the key players who were 'in the trenches' the whole time. As a consequence, there is some duplication of information as each contributor tells it from his or her own perspective. To have edited out this duplication would have destroyed the essential personal touch brought by each writer. All of their stories are buttressed by spectacular images and other data returned by NEAR Shoemaker.

We are particularly grateful to the individuals who accepted our invitation to contribute to this volume, but we would also like to acknowledge all of the scientists, engineers, managers, and support staff who worked together to make the NEAR mission such a success. It is impossible to name everyone involved, so instead we focus on the various teams that worked together to get the job done, day in and day out.

It all started with the Johns Hopkins University's Applied Physics Laboratory (APL) mission design team, who conceived the cruise path to Eros, which included the Mathilde flyby and the first use of an Earth gravity assist to change the orbital inclination of a planetary spacecraft. The spacecraft design and engineering team at APL and a number of other aerospace industry contractors built and tested the spacecraft and instruments and successfully delivered them to the Cape Canaveral launch facility ahead of schedule and under budget. Boeing and the Delta launch team at the Cape provided NEAR Shoemaker with a flawless ride into space. The navigation team at Caltech's Jet Propulsion Laboratory (JPL) provided the critical data and analyses needed to get NEAR from place to place precisely, including the painstaking task of examining thousands of images of Eros to identify and track landmarks needed to continually update and maintain the spacecraft's fragile orbit around the asteroid and then, amazingly, to execute a precision soft-landing at the end of the mission. The mission operations team at APL had the unenviable task of organizing all of the communications between the spacecraft and the NASA Deep Space Network, including uplinking of all of the instrument and spacecraft operation sequences and downlinking all of the science data and engineering telemetry. This they did with enthusiasm and skill, always maintaining the perfect balance of conservatism required for mission safety and the desire to maximize the scientific return from the mission as a whole. And, last but not least,

the science team, which was deeply involved in the daily operations of the spacecraft, worked painstakingly to design complex instrument observing sequences, to analyze the returned data quickly so as to influence future operations, and to publicize the results quickly in the scientific literature and, with APL's help, on the Web (primarily through the popular NEAR home page at http://near.jhuapl.edu).

The members of these teams are the real heroes of NEAR. Often working late hours, weekends, and holidays, often overcoming adversity and the constant pressures of space flight operations, and always maintaining constant lines of communications among each other and with the general public, they ultimately were responsible for making the mission a truly successful piece of the growing history of space exploration.

Jim Bell and Jacqueline Mitton

PICTURE CREDITS
Credit for images and other graphics in the book goes jointly to NASA and the Applied Physics Laboratory (APL) of Johns Hopkins University (JHU), with the following exceptions:

1.1 Nice Observatory.
1.2 J. Veverka.
1.3 J. Bell, generated with Voyager II software (Carina Software Inc.)
1.4 STScI and AAO.
1.8 Reproduced with permission, from J. Veverka *et al.*, *Science*, vol. 289, p. 2088. © *Science* 2000.
2.1, 2.2, 2.3, 2.4, 2.5, 2.6, 2.7, 2.8, 2.9, 2.10, 2.11, 2.12, 3.4, 3.9, 3.10, 3.15 JHU/APL.
4.3, 4.4 , 4.6 M. Robinson.
4.5 L. Prockter and M. Robinson.
4.14 NASA, JHU/APL and L. Prockter.
5.1, 5.3, 5.4 , 5.13 P. Thomas.
5.2 P. Thomas and M. Robinson.
5.7, 7.6 P. Thomas and J. Joseph.
6.1 Painting by Don Davis.
6.2 J. Bell, based on data from M. J. Gaffey, J. F. Bell, D. P. Cruikshank and R. Norton.
6.3 J. Bell, based on data and analyses by D. Tholen, J. Gradie and E. Tedesco.
6.4 NASA/JPL.
6.5 © E. Johnson and L. McFadden, University of Maryland. Used with permission.
6.9, 6.11, 6.14, 8.7 J. Bell.
6.12 A. Cheng.
6.17 S. Murchie.

7.1 (right) University of Arizona Press.
7.3 NASA, JPL, JHU/APL and P. Thomas.
7.7 C. Chapman.
7.10 NASA, JHU/APL and J. Bell.
8.1 Cornell University.
8.2, 8.13 (bottom) Provided through the courtesy of the Jet Propulsion Laboratory, California Institute of Technology, Pasadena, California.
8.4 NASA/JPL.

8.6 JPL/Caltech and B. Owen.
8.8, 8.9, 8.12 M. Bell, NASA and JHU/APL.
9.12 E. McCartney.

ACKNOWLEDGEMENT

Part of the research described in Chapter 8 was carried out at the Jet Propulsion Laboratory, California Institute of Technology, under a contract with the National Aeronautics and Space Administration.

Foreword

Carolyn Shoemaker
US Geological Survey

Interest in asteroids, the first of which, Ceres, was discovered 200 years ago in 1801, has gradually escalated in the last 20 years. The realization of their importance in solar system history has inspired searches to discover as many of these small planets as possible and to determine their orbits and physical properties. Today we understand that asteroids are primitive bodies, representative of the early planetesimals which collided to help form our large planets. We know that they continue to play a role in impact and crater formation on every solid body in our solar system. The awareness that asteroids may have been both a cause for mass extinction of life as well as a potential source for its origin has done much to generate public and scientific interest in knowing more about them. How do these objects affect mankind and Earth? How many are there, where do they reside in space, what are their constituents and shape, size, and colors, what is their history and future, what was their role in the origin of our solar system? Are they only an environmental hazard, or do they offer a stepping stone and an economic potential which could truly make our exploration of space something more than science fiction?

These questions and others spurred the remarkable journey of a small spacecraft from the Earth, past the large main-belt asteroid Mathilde, and on to an encounter with a Mars-crossing, near Earth asteroid named Eros. Launched by NASA on February 17, 1996, the spacecraft was the focus of the five-year NEAR (Near Earth Asteroid Rendezvous) mission to learn as much as possible about asteroids. A mission comprises much more than a spacecraft, of course, and its achievements depend on the expertise provided by engineers, administrators, and scientists. The success of the NEAR mission was due to many people, but the superb results have meaning also for many who were not directly involved with this flight.

Gene Shoemaker had the opportunity to take part in some of the early planning for NEAR as a member of the 1985 working group that studied objectives and design for this first mission to an asteroid. The planning stages for NEAR took a number of years, but eventually it became the first of the Discovery Series missions. Gene was enormously excited and pleased to watch the images of Mathilde, sent to Earth as the spacecraft executed its flyby on June 27, 1997. While studying the role of minor planets in the process of impact and cratering throughout our solar system, he had long dreamed of a mission to an asteroid.

Gene's was the approach of a scientist, a geologist. My interest in asteroids developed out of a sense of wonderment about these minor planets and their significance, a curiosity that arose from my desire to learn more about our astronomical neighborhood. A search program for these large rocks in the sky by means of a telescope became important to us both, and the thrill of discovery and learning something new was for us the carrot at the end of the stick. Throughout the years of our observing program, we often wondered why so few others were interested in asteroids and particularly in near Earth asteroids with all their implications and possibilities. "For everything there is a season," and the season seemed finally to be upon us with the discovery of Comet Shoemaker–Levy 9 and its collision with Jupiter. If comets collide with planets, why not asteroids?

Gene would have liked to send a manned mission to an asteroid, a minor planet. He felt that this was both feasible and economical. The energy required to send astronauts to an asteroid and back was less than to send them to the Moon or any other planet. A manned mission would have taught us more about survival in space as preparation for the exploration of Mars. It would have revealed the kind of information that the human brain is best suited to synthesize; a man with a rock hammer could read clues for which no instrument has yet been designed. Gene felt that if mankind is to venture into space, there must be a reason behind the venture. To accomplish very much, our astronauts need experience on other worlds, and asteroids offer that opportunity.

Gene also believed in unmanned missions as a way to probe, study, and learn about planets of all sizes, major and minor. He believed strongly in NEAR and knew from the 1994 Clementine mission that spacecraft instruments could yield a vast amount of information. Sadly, he did not live to see the superb results sent by this tiny spacecraft during its year in orbit about Eros, nor to see the conclusion with its amazing touchdown on Eros on February 12, 2001. Because of the part he played as a pioneer in interplanetary science and his influence on decades of asteroid and comet research, the successful entry of this craft into orbit about Eros prompted its renaming to "NEAR Shoemaker." This honor has special meaning to all who shared his passion for asteroid exploration. The conclusion of the mission was the answer to Gene's dream.

Prior to NEAR, telescopic observations had already told us much about asteroids, so-called because they appear "star-like" as points of reflected light when observed at the telescope. We knew them to be rocky bodies, mountains in the sky, with varying orbits, different densities, and a variety of compositions. There may be a cross-over between asteroids and comets, depending on the amount of ice and rock as constituents. These small bodies are found between Mars and Jupiter if they are in the main belt, or beyond Neptune if they are part of the Kuiper belt. As more and more asteroids have been discovered, our statistics on planet-crossing bodies have improved. Despite the enormous distances in space, it is possible that most main-belt asteroids have suffered collisions with each other. In many cases, this has set fragments on unusual orbits which could lead to planet crossing. It becomes possible for these rogue bodies to impact the terrestrial planets and Jupiter and its moons. Those that approach Earth are termed near Earth asteroids or, more generally, near Earth objects (NEOs). Eros is one of the first NEOs to have been discovered, the only one to date to be orbited by a man-made spacecraft and to be thoroughly studied, and the only one on which a landing has been attempted and successfully executed.

Toward the end of June in 1997, NEAR Shoemaker streaked past the asteroid 253 Mathilde, at a distance of 753 miles and a speed of 22 000 miles per hour. The images obtained there showed us that Mathilde has been impacted a number of times and would surely have been broken apart by collisions during formation of its largest craters, if it were a solid, dense body. Instead, measurements of its mass tell us that it must be made of many fragments held together by their mutual gravity and thus able to absorb the energy of a large impact without splitting or fragmenting.

On December 20, 1998, a computer glitch shut down the NEAR Shoemaker spacecraft and allowed it to tumble 2000 miles off course. Three days later, control was regained when the craft called home, and it was navigated back on course. The close encounter with Eros was delayed about a year but was still possible because of a contingency plan. Almost four years after leaving home, NEAR entered orbit around Eros and was renamed.

During the year spent in orbit, NEAR Shoemaker circled Eros 207 miles above the surface and gradually descended, with two passes at less than two miles above the poles. A variety of instruments, which included a magnetometer to look for a magnetic field, a multispectral imager to study rock types and geology, an infrared spectrograph to map mineral composition, a laser range finder to tell us about the shape of the asteroid, an X-ray sensor and gamma-ray sensor to measure key elements, and a radio science experiment to determine mass and density, fed continual data to the scientists on Earth. Every square inch of Eros came under study both from high orbit and low orbit just three miles above the surface. This was a challenging exercise, as this object is a low-mass irregularly shaped body and its gravitational grip is not firm.

NEAR Shoemaker provided details of rocks and soil, boulders, ridges, grooves, and craters. We learned that Eros has a uniform density and is not a rubble pile – it is solid. The chemical constituents of Eros are those discerned in all the planets. This little planet provides a cornerstone to our understanding of the primitive solar system, because it is essentially unaltered from its original state. As with all missions, there have been surprises along the way, and there will doubtless be more as the data are analyzed and studied by the scientists. Magnification of planetary bodies by close approach or landing provides new and unexpected information, as in the case of the Eros regolith, the numerous boulders on the surface, the apparent dust transportation, and lack of many small craters.

After five years in space, having completed its goals, NEAR Shoemaker descended to the surface in an astonishing, soft, controlled landing. Its elegant touchdown left the spacecraft still operating, sending signals and able to continue with the gamma-ray experiment.

This book, *Asteroid Rendezvous: NEAR Shoemaker's Adventures at Eros*, eloquently tells of our first attempt in this odyssey of scientific endeavor. It is a detailed, factual account of the exploration of an asteroid. The contributors draw upon a great deal of experience and expertise gained from many different missions, but they also recognize that the more we know, the more there is to know. One of the charms of space exploration is that there will always be new questions arising from the old. It is to be hoped that many different kinds of asteroids will be the targets of space missions, including manned landings, in the future, and that these bodies will become stepping stones, leading us on into our exploration of the solar system and beyond. NEAR Shoemaker has provided fundamental information in this quest.

July, 2001

1 Eros: Special among the Asteroids

Joseph Veverka
Cornell University (NEAR Camera Team leader)

"We will carry a telescope, in order to see Eros coming." With these words, Freeman Dyson, then nine years old, began a never-completed tale in 1933 called *Sir Phillip Robert's Erolunar Collision.* The story, an adventure in the genre of Jules Verne, describes an expedition to the Moon in order to witness a predicted collision between the asteroid Eros and our satellite! Ever since its discovery in 1898, Eros has been a notorious object and has attracted public attention – partly because it is a moderately large object and, astronomically speaking, can come fairly close to Earth.

Eros has always been a special asteroid, one connected with many firsts. The latest of its distinctions, and the focus of this book, is that it was the target for NASA's spectacularly successful NEAR (Near Earth Asteroid Rendezvous) mission, an endeavor that not only produced the first-ever global geological and geophysical characterization of an asteroid but also accomplished a successful landing on an asteroid with a spacecraft that was never designed to be anything more than an orbiter.

No ordinary asteroid

When discovered on August 13, 1898, independently by Gustav Witt, director of the Urania Observatory in Berlin, and by Auguste Charlois at the Observatory of Nice, it was the 433rd asteroid recorded, but soon it became evident that Eros was special. Its unusual orbit was quickly noted. On its significantly elongated elliptical path, it takes 1.76 years to make a complete circuit around the Sun and its distance from the Sun ranges between 1.78 and 1.33 Astronomical Units (AU). All previously discovered asteroids were confined to a belt between 2.0 and 3.3 AU from the Sun located between the orbits of Mars and Jupiter. Eros was the first asteroid known to cross the orbit of Mars, which has an average distance from the Sun of 1.52

AU. Eros's orbit is inclined to the main plane of the solar system by just under 11°. It was also the first asteroid known that can come fairly close to Earth. It can occasionally come as little as 0.15 AU (14 million miles) away. An encounter with Earth at this distance occurred in 1975, the closest Eros has been since its discovery.

An earlier close approach to Earth by Eros, in February 1931, was of particular interest to astronomers because it was the best opportunity that had ever arisen for determining the scale of distances in the solar system by means of parallax observations. Although the relative distances between the Sun and the planets were well known from Kepler's and Newton's laws, converting them into miles or kilometers relied on at least one absolute determination of the distance between Earth and a planet or asteroid. At that time, measuring a body's parallax was regarded as the principal technique for doing this. The nearer the object, the greater its parallax and, in principle, the easier it was to measure. No other body in orbit around the Sun had been observed to come closer than Eros, so an international observing campaign was launched. Quite separately, the average distance

Figure 1.1. A 1910 etching showing a general view of the Observatory of Nice, one of the two observatories where asteroid 433 Eros was discovered independently in 1898 (the other being the Urania Observatory in Berlin). The 24-m-diameter dome was built in 1881 and is still the largest telescope dome in Europe. It houses the famous 76-cm (30-inch) "Grande Lunette" telescope, which is 18 m (59 feet) long.

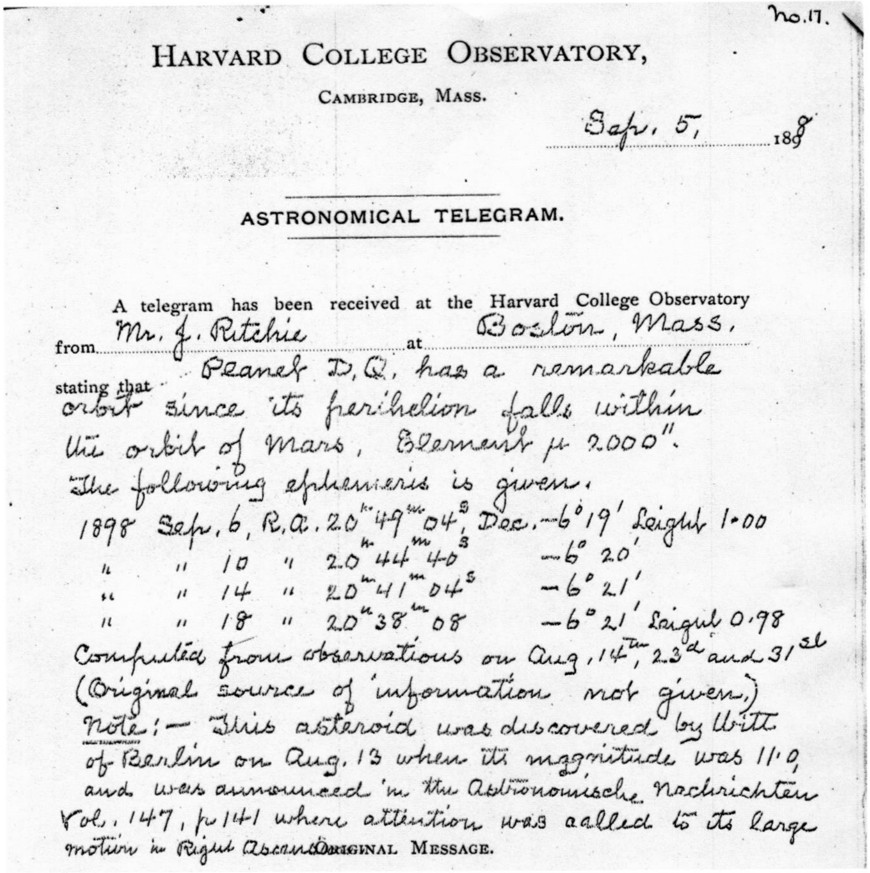

Figure 1.2. Telegrams from the Harvard College Observatory's archives announcing (top) the observation of the "remarkable orbit" of minor planet 1898 DQ, later re-named 433 Eros, by J. Ritchie, and (bottom) the discovery of substantial variations in the light curve of Eros by E. Pickering.

HARVARD COLLEGE OBSERVATORY,

CAMBRIDGE, MASS.

no. 17.

Sep. 5, 189

ASTRONOMICAL TELEGRAM.

A telegram has been received at the Harvard College Observatory from Mr. J. Ritchie at Boston, Mass. stating that Planet D.Q. has a remarkable orbit since its perihelion falls within the orbit of Mars, Element μ 2.000". The following ephemeris is given.

1898 Sep. 6, R.A. 20ʰ 49ᵐ 04ˢ, Dec. −6° 19′ Light 1.00
" " 10 " 20 44 40 −6 20′
" " 14 " 20 41 04 −6 21′
" " 18 " 20 38 08 −6 21′ Light 0.98

Computed from observations on Aug. 14th, 23d and 31st (Original source of information not given.)

Note:— This asteroid was discovered by Witt of Berlin on Aug. 13 when its magnitude was 11.0, and was announced in the Astronomische Nachrichten Vol. 147, p. 141 where attention was called to its large motion in Right Ascension.

ORIGINAL MESSAGE.

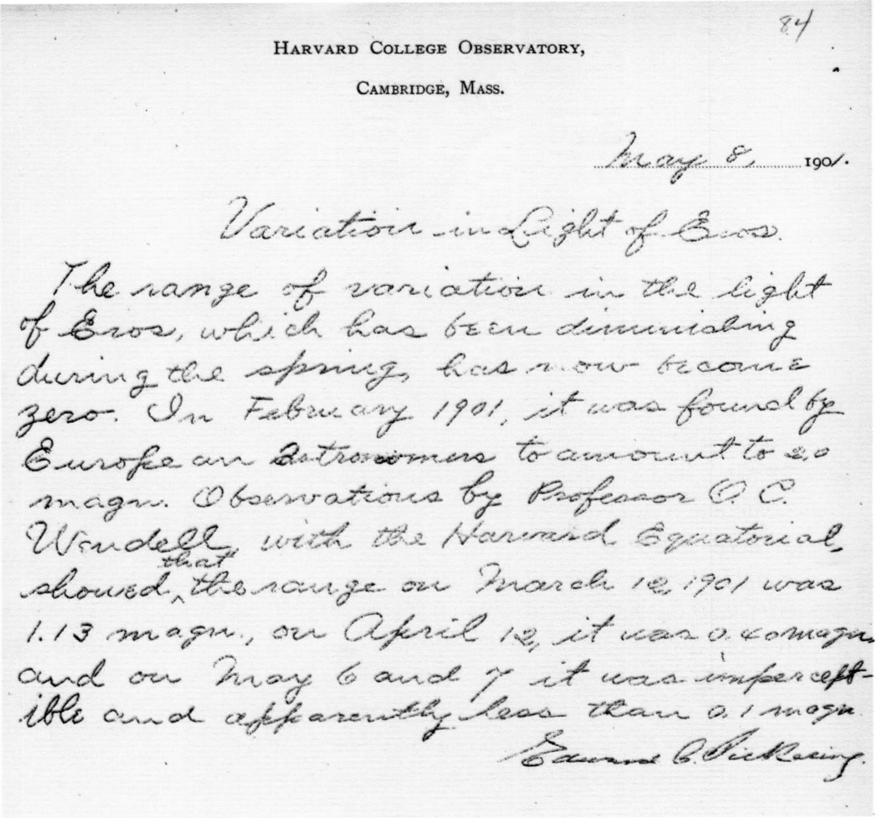

HARVARD COLLEGE OBSERVATORY,

CAMBRIDGE, MASS.

84

May 8 1901.

Variation in Light of Eros.

The range of variation in the light of Eros, which has been diminishing during the spring, has now become zero. In February 1901, it was found by European Astronomers to amount to 2.0 magn. Observations by Professor O.C. Wendell with the Harvard Equatorial, showed that the range on March 12, 1901 was 1.13 magn, on April 12, it was 0.6 magn, and on May 6 and 7 it was imperceptible and apparently less than 0.1 magn.

Edward C. Pickering

Figure 1.3. Top: A computer simulation of the positions of 5383 asteroids
on February 17, 1996, as viewed by an
observer far above the north pole of
the Sun. Red squares represent the
asteroids, and the colored circles represent the orbits of Mercury, Venus,
Earth, Mars, and Jupiter. Bottom:
View from the same vantage point
showing the orbits of the two asteroids studied by the NEAR mission,
433 Eros and 253 Mathilde.

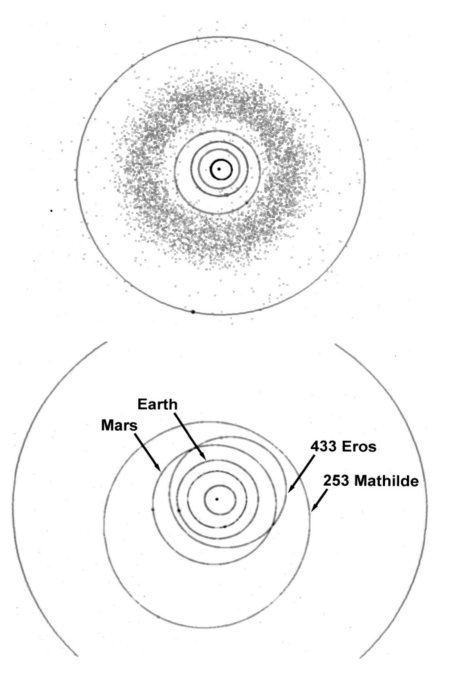

between Earth and the Sun – the astronomical unit –
was also calculated from observations of Eros by
analyzing perturbations to Eros's orbit due to the
gravitational influence of Earth and other planets.
Later, better methods based on radar superceded this
use of Eros, but humans, forever preoccupied with
cosmic destruction, have not forgotten that Eros has a
small but non-zero chance of one day hitting Earth or
the Moon.

Almost immediately after its discovery, it was noted
that Eros is also unusual due to its conspicuous and

periodic fluctuations in brightness, known to
astronomers as its "light curve". As could be expected,
these fluctuations received a variety of explanations,
including the suggestion that Eros is a binary asteroid,
before the correct interpretation was generally
accepted: Eros's brightness varies because it is an
irregularly shaped object rotating around a stable spin
axis. When at its closest, Eros reaches magnitude 7.2
and is bright enough to be seen with a small telescope
or binoculars, but it varies by as much as a magnitude
and a half over its rotation period of 5.3 hours.

Figure 1.4. Eros photographed in 1986 as part of the Anglo-Australian Observatory's southern sky survey. Since the telescope was tracking the stars, the faster-moving asteroid appeared as a streak in this long-exposure photograph. This is typically how asteroids have been discovered in photographic and digital plates, and the occasional bright asteroid seen streaking through fields of dim galaxies in painstaking long-exposure photographs is what earned asteroids their description "vermin of the skies" in the early 20th century.

During its close approach to Earth in 1931, Eros became the first asteroid (and probably still the only asteroid) to have its shape determined visually. In a tour-de-force observation, two observers in South Africa (van de Bos and Finsen) actually saw its elongated shape with the 67-cm (26.5-inch) refractor at the Union Observatory in Johannesburg and estimated an average diameter of about 23 km (14 miles), which is surprisingly close to the actual value of about 21 km. During the same period, observations of Eros served as a testing ground for new techniques that were being developed to determine the orientation of an asteroid's spin axis from light-curve records. These techniques, only slightly modified, are still in use today. In 1960, Eros became one of the first asteroids (actually the third) to be detected by radar and, in the 1970s, one of the first asteroids for which thermal measurements in the infrared strongly suggested that the topmost surface layer is not bare rock but, somewhat like the surface of the Moon, is covered by a thermally insulating blanket of pulverized and fragmented rock known as "regolith." Spectral reflectance measurements during this period showed Eros to be a member of the broad class of compositionally diverse S-type asteroids, which dominate the inner half of the asteroid belt (see Chapter 6).

More recently, Eros has become an object cherished by a thriving body of cottage industrialists who assiduously calculate impact hazards from Earth-approaching objects. Such calculations are probabilistic in nature. There is no way at present to calculate the long-term future of an object such as

Figure 1.5. Strong variations in the light curve of Eros as observed by ground-based telescopes are primarily caused by the asteroid's irregular shape. This large saddle-shaped depression was imaged by NEAR Shoemaker on March 22, 2000 from a range of 208 km. This feature was subsequently named Himeros, in honor of Eros's primary attendant in Greek mythology.

Eros with precision. Decades ago, Ernst Öpik showed that objects in orbits such as that of Eros have lifetimes of only one hundred million to one billion years before their courses are dramatically altered, usually by a close encounter (or even a collision) with a planet. More recent calculations have confirmed such estimates for Eros. Eros has not always been in its current orbit, and will not always be around for humans to worry about. What will happen to it in the long run cannot be predicted with certainty, however. Yes, it could possibly hit Earth (or the Moon) as was already clear to the nine-year-old Freeman Dyson in 1933. One thing that is more certain is that Eros appears to have come from the main asteroid belt and is only a "recent" interloper in near-Earth space.

Asking the right questions

The mission that eventually became NEAR was conceived in the mid 1980s. Looking back after its successful completion a decade and a half later, it is

Figure 1.6. Even before the first space-craft encounters with asteroids, astronomers knew that rapidly changing lighting conditions on these strangely shaped small bodies could drastically influence the ability to interpret their geology. An example is shown here, from NEAR images obtained on June 23, 2000 from an orbital altitude of 52 km (33 miles). Structural and topographic features are accentuated in the top row images taken just after local sunrise; these are washed out and intrinsic differences in albedo begin to dominate in the bottom row images, taken slightly later with the Sun higher in the sky.

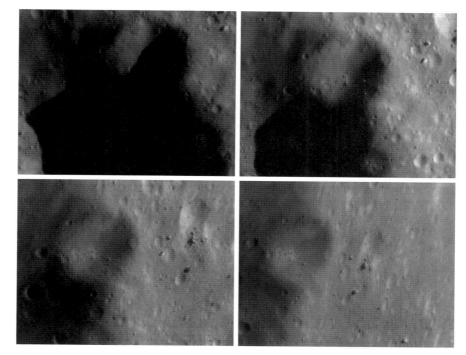

instructive to remind ourselves of the major questions that NEAR was designed to answer, recalling that, at the time when NEAR was first proposed, no asteroid had been visited by a spacecraft. The original questions were considerably refined in the decade before NEAR's launch in 1996, in view of the ever-growing store of information produced by assiduous telescopic and radar observations of asteroids and by the Galileo spacecraft's historic flybys of asteroids 951 Gaspra and 243 Ida in 1991 and 1993, respectively.

Shortly before the NEAR Shoemaker craft was to go into orbit around Eros, a press conference was held at NASA Headquarters in Washington. "What were the

Figure 1.7. One question astronomers wanted to address with the NEAR mission was what happens to all of the ejected material after a small body like Eros collides with other small bodies? Is it all thrown off the asteroid because of the low gravity, or does it disperse and collect on the surface over time? Images from NEAR like these show that a remarkable amount of impact debris has in fact accumulated on Eros over time. (Top) Image obtained on December 19, 2000 from an orbital altitude of 37 km (23 miles). This 1.5-km-wide region near the southern part of Himeros is littered with boulders. The largest boulder, near the center of the image, is about 60 m (200 feet) across. (Bottom) Image obtained on May 14, 2000 from an orbital altitude of 50 km (31 miles). This 1.8-km-wide region shows numerous rocks and boulders, many collected within the bowl of a small impact crater. The smallest rocks visible in this scene are only 4 m (13 feet) across; the largest boulders are about 60 m (200 feet) across.

high-resolution images, the spectral maps, and other anticipated data likely to reveal about Eros?" the media wanted to know. As always in science, getting an answer to a question is only part of the process. Asking the right question to begin with often proves equally important. What, then, were some of the major questions we were smart enough to ask at the time?

One important question concerned the record of past impacts on Eros that would be revealed by the detailed images of the craters on its surface. The known irregular shape of the asteroid pointed strongly to a vigorous collision history. But how violent had it really been? And, what clues would be preserved on the surface? Would the surface be uniformly and densely covered with craters? Would there be evidence of a catastrophic collision that had spalled off a large portion of the asteroid? Would there perhaps still be collision fragments in orbit as small satellites? How much of the surface would be covered with ejecta produced by the cratering of the surface? Given the very low gravity and miniscule escape velocity, how uniform and how thick would the regolith on the asteroid be?

More importantly perhaps, what did the collision history do to the interior of the asteroid? What is the inside of Eros like today? Is it perhaps a loosely bound conglomeration of rubble, as NEAR found asteroid 253 Mathilde to be when it measured its surprisingly low density during the June 1997 flyby? Is Eros perhaps a shard of strong, rigid, rock, a fortunate survivor of a catastrophic collisional trauma on its parent body? Or, is Eros something in between – a fractured shard, a body that started out the latest chapter of its evolution as a fairly strong and solid fragment but is gradually being weakened by fracturing associated with impacts?

A body in disguise?

Finally, there was the very important question, "What is Eros made of?" Is there a generally homogeneous make-up to it as a whole, or is Eros compositionally heterogeneous? Is the surface covered by a regolith, which may have a somewhat different composition from the interior? Are there holes in the regolith cover through which we can see the composition of the deeper levels? Are there ejecta blocks from large

Figure 1.8. The elongated, irregular shape of Eros does not provide unambiguous clues about the asteroid's interior structure and composition. To begin to categorize and interpret different geologic processes at work on a small body such as this, NEAR scientists had to define a coordinate system so that features could be located and intercompared. The science team established a latitude/longitude grid as a reference frame, choosing a large, bright crater (arrow) as the "prime meridian".

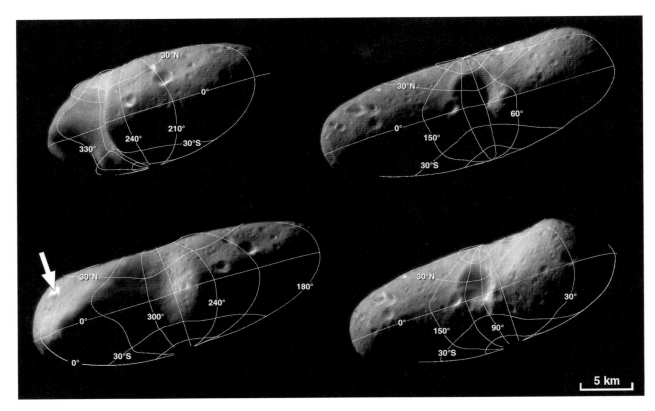

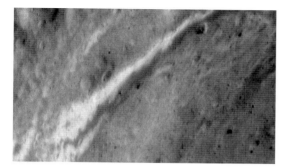

Figure 1.9. While some brightness variations on Eros are clearly related to the asteroid's shape, others appear related to variations in the intrinsic composition or physical properties of the surface itself. An example of such variation can be seen in this NEAR image of the inside of Himeros, obtained on May 30, 2000 from an orbital altitude of 49 km (30 miles). The scene is 1.8 km (1.2 miles) across. Topographic contrast is subdued and differences in reflectivity are enhanced because the Sun was high overhead when the image was taken. The curved feature extending from lower left to upper right is the inside of a broad, shallow groove. Bright material covers parts of the groove's walls. The boulder poised near the edge of the groove (upper right) is about 60 m (200 feet) across.

Figure 1.10. Loose material appears to litter the surface of Eros, sometimes even appearing to adhere to the surface at unnatural and counterintuitive angles, as in this stunning NEAR image obtained on June 20, 2000 from an orbital altitude of 51 km (32 miles).

craters that suggest that at depth the composition of Eros is different from what we see higher up? In short, what are the rocks on the inside of Eros like and how similar are they to the minerals in the upper veneer of regolith? How is the composition related to that of known meteorites, especially the most common class of meteorites, the ordinary chondrites? Establishing what links may exist with meteorites is crucial since it would then be possible to say something more about the evolution of bodies such as Eros from the detailed geochemical analyses that have been performed on related meteorites.

A major aim of NEAR was to try to resolve the "dichotomy of origins" of the S-type asteroids. S-type asteroids are like disguised objects – things that superficially look very similar in terms of their spectra but could be very different underneath. As a consequence, the class encompasses a range of fundamentally different kinds of asteroids. At one extreme are very primitive bodies for which cratering and collisions have been the dominant evolutionary processes and, at the other, are fragments of partially or wholly differentiated bodies. In either extreme case the exteriors can have S-type spectra, as has been demonstrated by Michael Gaffey and his coworkers. By getting close and making additional measurements, holes in the disguise might be detected through which the asteroid's true nature would be revealed. To mention just one simple example: if at small scales significant mineralogical diversity were found, this fact would indicate

unambiguously that Eros is a fragment of a differentiated body. If no such diversity were found the conclusion would be more ambiguous. It might mean that one was dealing with a body of uniform composition throughout, as the regolith indicated, or one could argue that nature had been particularly perverse and had produced a well-mixed outer regolith, which represented the average make-up of Eros, but masked the diversity that was present underneath. In the latter case, additional information would be needed to resolve the ambiguity. For example, if the distribution of mass within Eros could be measured accurately and the asteroid turned out to be highly homogeneous, this would be powerful evidence that no significant compositional difference lurks underneath the uniform regolith and that indeed Eros is, as it seems from the outside, a piece of an undifferentiated object.

One important question about smaller asteroids such as Eros that NEAR resolved in a dramatic fashion is that these objects, in spite of their miniscule gravity and low escape velocity, do retain enough of the debris excavated during collisions to have a surface covered by regolith, which in places may be quite deep and

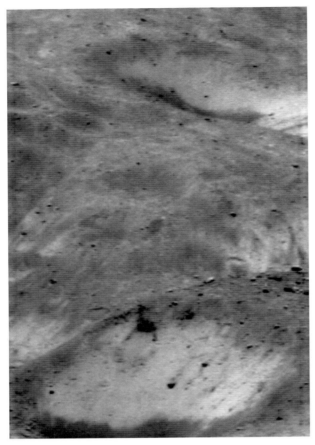

Figure 1.11. Evidence for a regolith – a blanket of pulverized and fragmented rock covering Eros – comes from NEAR images like this one, obtained on July 19, 2000 from an orbital altitude of 36 km (22 miles). The surface is pockmarked with small impact depressions, littered with rock fragments and boulders, and shows striations that provide evidence for downslope motion of regolith materials. The image shows a region about 800 m (2600 feet) across. The size of the smallest visible rocks is only about 6 m (19 feet).

stratigraphically complex. Manifestations of abundant regolith are evident in most high-resolution views of Eros. Perhaps the most striking (especially if one spent much time on the surface!) are the abundant ejecta blocks that litter the landscape. But more subtle signs are also present, ranging from extensive debris slumps on crater walls and other slopes to pond-like accumulations of presumably finer-grained regolith in crater bottoms and other local gravitational lows. Interpretations drawn from thermal measurements made telescopically indicating that the surface of Eros has the thermal inertia of a regolith, have been borne out, but the extent of the regolith and the complexity of the regolith process documented in the NEAR images have come as a surprise to most investigators.

A benchmark for asteroid studies

Now that we have a complete global characterization of Eros in hand it is also valuable to ask how well inferences based on pre-NEAR studies of the asteroids have held up. This test provides a benchmark for how much faith we should have in the information on hundreds of asteroids studied from the ground but, as yet, not visited by spacecraft.

Eros's spin period and the direction of its spin axis were determined reliably from light-curve observations even in the 1930s. The maximum amplitude of the light curve yielded a good estimate of the ratio between the lengths of the principal axes of Eros, but the absolute size of the object caused more of a problem (in spite of the observations of van de Bos and Finsen). Much of the difficulty in establishing its absolute size was due to not knowing its reflectance or albedo. The situation improved in the 1970s when estimates of the albedo were obtained independently by infrared radiometry and polarimetry. These and subsequent estimates varied considerably in the case of Eros for reasons that are not obvious, even in retrospect. The radiometry, as noted above, did provide positive evidence of a regolith. Radar observations of Eros led to the determination of a shape – a "convex hull" – which closely approximates the banana shape of the asteroid found by NEAR. Interestingly, although the shape was determined well, the best estimate of the absolute dimension prior to NEAR was some 40.5 by 14.5 km, somewhat bigger than the true size. With few exceptions, most observers did not detect significant color variations in the light from the whole disk of Eros as it spins, consistent with NEAR's finding of an almost homogeneous regolith across the body.

Even though the NEAR mission is now over, astronomers and other scientists will continue to draw new conclusions from its great harvest of data for some time to come and will assess how our scientific conception of asteroids and of processes that dictate their evolution has changed in the light of the results obtained by NEAR. Later chapters in this book summarize what was found out during the course of the mission and in the first few months after it ended, and the new quandries the data raise.

Great progress has been made. A century ago Eros was only a point of light in the sky avidly studied by astronomers because of its unusual orbit and puzzling brightness variations. Now, thanks to NEAR, Eros has become one of the best-known small worlds in the

Figure 1.12. Some of the NEAR multispectral imager/Near infrared spectrometer science team, and their associates, ponder the mysteries of Eros, Mathilde, and other small bodies during a team meeting at APL in June, 1997. From left to right, are: Joe Veverka (team leader), Jonathan Joseph, Peter Thomas, Carolyn Porco (visiting guest), Bill Merline, Clark Chapman, Jim Bell (seated), Scott Murchie (standing), Beth Clark, Mike Malin, Mark Robinson, and Jesse and Jeremy Veverka (guests).

solar system, and one that will richly repay future exploration with rovers and sample return vehicles. If an orbiter can land on an asteroid as NEAR Shoemaker did in February 2001, landing a properly designed rover should be feasible. NEAR has shown that even asteroids as small as Eros display a wide diversity of surface features and processes, many of which merit closer investigation. The evidence gathered by NEAR, that subtle compositional differences do occur at different depths in the regolith, calls for detailed in-situ compositional analysis. NEAR has shown that the gathering of samples on Eros should be easy: there are plenty of rocks of all sizes and an abundance of more finely textured regolith in most places.

Eros, an asteroid of firsts, and NEAR, a mission of firsts, have come together in an amazing adventure that has underscored how fascinating asteroids are to explore. Their exploration will improve our understanding of the processes that have affected the evolution of these remnants of the building blocks of the

Earth-like planets. The success of NEAR at Eros should be followed by a methodical effort to explore other asteroids – objects of different types, different sizes, and at different locations in the solar system. But equally important to the advancement of asteroid science in the 21st century will be a return to Eros with a well-equipped rover and a sample-return spacecraft to resolve some of the mysteries about the evolution of S-type asteroid regoliths that NEAR has left us.

"Three cheers for Eros!"

(F. Dyson, *Sir Phillip Robert's Erolunar Collision*, 1933)

2 A Date with Eros

Robert Farquhar
Applied Physics Laboratory, Johns Hopkins University (NEAR Mission Director)

Designing the mission

In 1983, NASA's Solar System Exploration committee suggested that a rendezvous would be an ideal way to begin a program of spacecraft missions to near Earth asteroids. This was followed by a comprehensive assessment of a possible Near Earth Asteroid Rendezvous (NEAR) mission by a distinguished Science Working Group (SWG) in 1986. The SWG found that the NEAR mission could be done with a Planetary Observer spacecraft, and also concluded that almost any near Earth asteroid in an accessible orbit would be a suitable target for the first mission.

Unfortunately, though, NASA's Planetary Observer Program was terminated after only one mission, the ill-fated Mars Observer, so plans for a NEAR mission remained in limbo after the 1986 report. However, in 1990, NASA introduced a new program of low-cost planetary missions called "Discovery." A Discovery Science Working Group (DSWG) was established and in October 1991 it recommended that "the first mission of the Discovery Program should be a rendezvous with a near Earth asteroid."

In 1991, the Applied Physics Laboratory (APL) of Johns Hopkins University and the Jet Propulsion Laboratory (JPL) of Caltech prepared competitive proposals for the NEAR mission. After a thorough review of the two proposals by a select panel of experienced project managers, NASA awarded primary management responsibility for the mission to the APL team. APL carried out system definition studies in 1992–1993. Construction of the spacecraft and instruments began in December 1993, and the spacecraft was shipped to the Kennedy Space Center about two years later.

The original plan called for a launch in January 1998 to the near Earth asteroid 4660 Nereus. Nereus satisfied the minimum requirements of the DSWG for a first mission to a near Earth asteroid, but its diameter

is only about 1 km (half a mile). Some scientists were concerned that its small size would limit the quantity and diversity of scientific results from the mission, so planners tried to find some way to reach a larger near Earth asteroid. 433 Eros, one of the largest and most accessible near Earth asteroids, quickly became the preferred target.

In some of the earliest proposals for a mission to a near Earth asteroid, Eros had frequently been the "target of choice." The disadvantage was that a rendezvous with Eros takes a lot of energy. It requires a launch on a trajectory that is highly inclined to Earth's equator, which does not receive the full benefit of Earth's rotation. However, in September 1992, NEAR mission designers came up with a trajectory that would significantly reduce the energy requirement. The method is called ΔVEGA – short for ΔV and Earth Gravity Assist. ("Delta-V" is the change in a spacecraft's velocity.)

The flight was to begin with a launch in February 1996. A deep-space maneuver on July 3, 1997, would reduce the perihelion distance of the spacecraft's trajectory, which would have the effect of increasing the spacecraft's energy when it swung by Earth in January 1998. In a new application of the ΔVEGA technique, the additional energy would be channeled not into increasing the aphelion of the spacecraft trajectory but into changing the inclination of its orbit relative to the plane of Earth's orbit by approximately 10°. Following the Earth-swingby maneuver, NEAR was to arrive at Eros on January 10, 1999. The circuitous three-year flight path to Eros resulted from a Discovery Program requirement to use an inexpensive, but less-capable, Delta-2 launch vehicle. With a larger launch vehicle, such as an Atlas or Titan, a one-year direct trajectory would have been possible, but the total mission cost would have increased by at least $50 million.

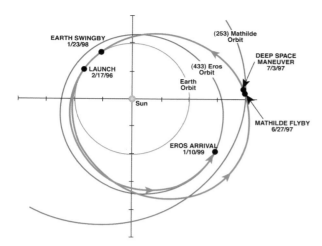

Figure 2.1. NEAR Shoemaker's planned trajectory through the inner solar system. The spacecraft was launched on February 17, 1996 and first traveled out beyond the orbit of Mars and into the main asteroid belt, flying past the C-type asteroid 253 Mathilde on June 27, 1997. NEAR Shoemaker then came back to Earth and used Earth's gravity as a slingshot on to Eros. The nominal mission profile would have had NEAR Shoemaker arrive at Eros on January 10, 1999.

In addition to the cost savings, the indirect ΔVEGA trajectory offered the possibility of encountering a main-belt asteroid on the way to Eros as the spacecraft passed through the inner portion of the main asteroid belt. In late 1994, NEAR's planned path was examined to see if it would come close to any interesting asteroid or comet. Fortuitously, NEAR's trajectory would pass within 0.015 Astronomical Units (AU), or only about 2.3 million km (1.4 million miles) of the large main-belt asteroid 253 Mathilde. Of course, a deviation of 0.015 AU from the nominal trajectory was not inconsequential, but further analysis revealed that the change was achievable. However, an encounter with Mathilde at roughly 2 AU from the Sun, twice Earth's distance from the Sun, was not without risk. Adding the Mathilde flyby would further complicate an already complex mission plan. NEAR was custom-made to orbit Eros, not to execute an asteroid flyby under circumstances where the available solar power would be considerably less than planned for. Worse yet, some mission engineers and astronomers worried that dust particles orbiting Mathilde might destroy the spacecraft as it flew past at 10 km/s (22 500 miles per hour). Many project personnel, including the project scientist, argued that the risks were too high and that the prudent thing to do was to ignore the possible flyby of Mathilde. In the end, however, the argument that important and unique science could be obtained

from a Mathilde flyby carried the day, and NEAR's trajectory was modified to include an encounter with Mathilde on June 27, 1997.

The spacecraft

Simplicity and low cost were the main principles behind the design of the spacecraft, which since it entered orbit around Eros has been known as NEAR Shoemaker (while the enterprise as a whole is the NEAR mission). Simplicity was achieved by having all three major components, the instruments, solar panels, and high-gain antenna, fixed and mounted on the body of the spacecraft. Although this requirement would increase the complexity of operations to some extent, it was an important factor in keeping the expense down. NEAR's development cost was less than $115 million in real-year dollars, well under the cap for a Discovery mission of $150 million (in financial year 1992 dollars). Also the development phase lasted only 26 months, significantly less than the upper limit of 35 months for a Discovery mission.

The spacecraft was three-axis stabilized using four reaction wheels to control pointing: each wheel could be sped up or slowed down to impart momentum that would gently roll the spacecraft along a desired axis. It had four large gallium-arsenide solar panels that could provide 350 watts of power even at NEAR Shoemaker's maximum solar distance of 2.2 AU. In addition to the 1.5-m high-gain antenna, there were two low-gain antennas and a medium-gain antenna for slower but more reliable communications. Data rates during the rendezvous phase were quite respectable, and could be as high as 26.5 kilobits per second when using the 70-m Deep Space Network (DSN) antennas. NEAR Shoemaker could also store as much as 1.7 gigabits of data on its two solid-state recorders. The dual-mode propulsion system consisted of one large 450-newton thruster and 11 smaller thrusters. The large engine used a combination of two propellants – hydrazine and nitrogen tetroxide – while the smaller thrusters used pure hydrazine.

The science payload consisted of five instruments (or six if the X-ray/gamma-ray spectrometer, which was two instruments in one, is considered as two):

1 Multispectral imager (MSI)
2 Near-infrared spectrometer (NIS)
3 X-ray/gamma-ray spectrometer (XGRS)
4 NEAR laser rangefinder (NLR)
5 Magnetometer (MAG)

Figure 2.2. A summary of the properties of the NEAR Shoemaker spacecraft.

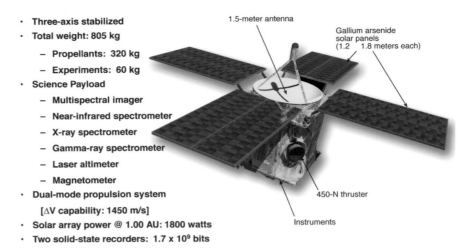

- Three-axis stabilized
- Total weight: 805 kg
  - Propellants: 320 kg
  - Experiments: 60 kg
- Science Payload
  - Multispectral imager
  - Near-infrared spectrometer
  - X-ray spectrometer
  - Gamma-ray spectrometer
  - Laser altimeter
  - Magnetometer
- Dual-mode propulsion system
  [ΔV capability: 1450 m/s]
- Solar array power @ 1.00 AU: 1800 watts
- Two solid-state recorders: 1.7 x 10⁹ bits

Figure 2.3. The NEAR mission scientific instruments. The imaging and
spectroscopy instruments were
mounted on the spacecraft's lower
deck, so that they could point at the
asteroid while the solar panels
remained directed towards the Sun.
The magnetometer, however, was
placed on the "quiet" topside of the
spacecraft in order to maximize its
sensitivity. The X-ray solar monitor
was also placed near the top to allow
a clear view of the Sun.

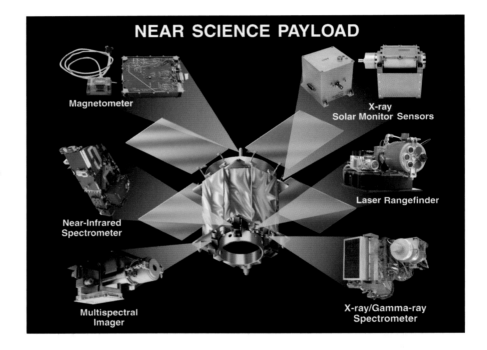

The MAG was mounted on top of the high-gain
antenna feed, where it is exposed to the minimum
level of spacecraft-generated magnetic fields. The
remaining instruments were all mounted on the
outside of the aft deck. They were on fixed mounts and
were co-aligned to view a common boresight direction. The NIS had a scan mirror that allowed it to look
more than 90° away from the common boresight.

Multispectral imager
The main goals of the MSI were to determine the shape
of Eros and to map the mineralogy and morphology of
features on its surface at high spatial resolution. The
charge-coupled device (CCD) camera had five-element,
radiation-hardened, refractive optics. It was sensitive

to wavelengths between 0.4 and 1.1 microns, and the
CCD was an array of 537×244 pixels. An eight-
position filter wheel contained filters chosen to
optimize the MSI's ability to detect minerals expected
or known to occur on Eros. The camera had a field of
view of 2.90 by 2.25° and a pixel resolution of 95×161
milliradians. It could take frames at a maximum rate of
one per second with images digitized to 12 bits and had
its own dedicated digital processing unit.

Near-infrared spectrometer
The NIS was designed to measure the spectrum of sunlight reflected from Eros in 64 near-infrared channels
at wavelengths in the range 0.8–2.6 microns. The ultimate purpose was to determine the distribution and

abundance of surface minerals, such as olivine and pyroxene. This grating spectrometer dispersed the light from the slit across a pair of passively cooled one-dimensional array detectors; one a germanium array covering the shorter wavelengths 0.8–1.5 microns and the other an indium/gallium-arsenide array covering 1.3–2.6 microns. The NIS had a scan mirror that enabled it to view the common boresight direction (within the field of view of the MSI) or directions more than 90° away. Spectral images could be built up by a combination of scan mirror and space-craft motions. The NIS had a gold calibration target that could be positioned to scatter sunlight into the instrument for in-flight calibration.

X-ray spectrometer

The XRS was an X-ray resonance fluorescence spec-trometer that detected the characteristic X-ray line emissions caused by solar X-rays impinging on major elements in the asteroid's surface, including magne-sium, aluminum, silicon, calcium, and iron. It covered X-rays in the energy range 1–10 keV using three gas proportional counters. Its field of view was 5°, allow-ing it to map the chemical composition of the asteroid at spatial resolutions as great as 2 km in the low orbits. A separate solar monitor system continuously mea-sured the incident spectrum of solar X-rays, using both a gas proportional counter and a high-spectral-resolution silicon X-ray detector. The XRS performed in-flight calibration using a rod with ^{55}Fe sources that could be rotated into or out of the detector's field of view.

Gamma-ray spectrometer

The GRS was designed to detect characteristic gamma rays in the energy range 0.3–10 MeV range that are emitted from specific elements in the surface. Some of these emissions are excited by cosmic rays and some arise from natural radioactivity in the asteroid. The GRS used a body-mounted, passively cooled, NaI scin-tillator detector with a bismuth germanate anti-coincidence shield to define a 45° field of view.

NEAR laser rangefinder

The NLR was a laser altimeter that measured the dis-tance from the spacecraft to the asteroid surface by sending out a short burst of laser light and then recording the time required for the signal to bounce back. It used a chromium/neodymium/yttrium-alu-minum-garnet (Cr-Nd-YAG) solid-state laser and a compact reflecting telescope. It sent a small portion of

each emitted laser pulse through an optical fiber of known length and into the receiver, providing a con-tinuous in-flight calibration of the timing circuit. The ranging data were acquired in order to construct a global shape model and a global topographic map of Eros with a horizontal resolution of about 300 m. The NLR also measured detailed topographic profiles of surface features on Eros with a best spatial resolution of about 5 m.

Magnetometer

The fluxgate magnetometer used ring core sensors made of highly magnetically permeable material. It was intended to search for and map the intrinsic mag-netic field of Eros.

Radio science

In addition to the six major instruments, a coherent X-band transponder was used to conduct a radio science investigation by measuring the Doppler shift due to the spacecraft's radial velocity relative to Earth. Accurate measurements of the Doppler shift as the spacecraft orbited Eros were used to map the asteroid's gravity field.

Journey to Eros

NEAR Shoemaker's journey to Eros began on February 17, 1996 when the spacecraft was successfully launched by a Delta-2 rocket. The launch was close to nominal, requiring only two trajectory correction maneuvers to remove the injection errors. The total ΔV was 11.5 m/s. On February 18, 1997, NEAR Shoemaker reached its aphelion distance of 2.18 AU, setting a new record for the distance from the Sun at which a space-craft is powered by solar cells.

The first major event in NEAR Shoemaker's flight was the flyby of 253 Mathilde on June 27, 1997. Although Mathilde was discovered in 1885, it was only after the announcement of NEAR Shoemaker's planned flyby that extensive telescopic observations were initiated. These observations revealed that Mathilde is one of the slowest-spinning asteroids, taking 17.4 days to complete one rotation. It was also found that Mathilde's spectrum was consistent with those of dark, carbon-rich, C-class asteroids. This last fact was important because the only previous space-craft encounters with asteroids were the Galileo flybys of 951 Gaspra and 243 Ida, which are both brighter, silicate-rich S-class asteroids.

NEAR Shoemaker's encounter with Mathilde was

Robert Farquhar

Figure 2.4. Launch of the NEAR
mission on February 17, 1996 from
Cape Canaveral, Florida. The spacecraft was nestled in the upper nose
cone section of the Delta-2 rocket #232.

Figure 2.5. Geometry of NEAR Shoemaker's Earth swingby encounter on January 23, 1998. The spacecraft approached Earth from high northern latitudes and made its closest approach to Earth, 540 km (335 miles) above southwestern Iran. Earth's gravity was used to bend the path of the spacecraft into a more inclined orbit, matching it to the orbital tilt of Eros. Universal time ticks are shown for reference.

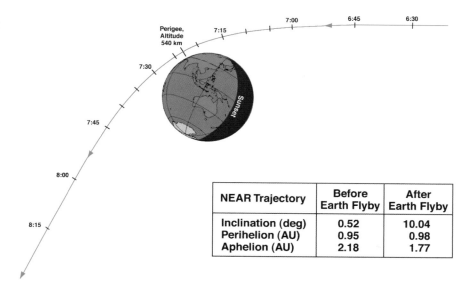

NEAR Trajectory	Before Earth Flyby	After Earth Flyby
Inclination (deg)	0.52	10.04
Perihelion (AU)	0.95	0.98
Aphelion (AU)	2.18	1.77

unusually difficult for a number of reasons. First, the spacecraft approached Mathilde from its night side, at a phase angle of 140°. From the point of view of the spacecraft, only a small part of Mathilde was illuminated by the Sun and the asteroid was only 40° away from the Sun. This created a severe problem for obtaining optical navigation images like those described in Chapter 8. Mathilde was first detected just 36 hours before closest approach as a faint dot almost lost in the Sun's glare. Second, the encounter took place at about 2 AU from the Sun where the power available from the spacecraft's solar panels was only 25% of the maximum for the mission. Furthermore, because the entire spacecraft had to be turned to point the camera at Mathilde, it was necessary to orient the solar panels about 50° away from the optimal solar direction during the encounter, which reduced the available power by another 36%. To conserve power, only one of NEAR Shoemaker's six instruments, the multispectral imager, was on during the encounter. Finally, because NEAR Shoemaker did not have a scan platform for the camera, the design of the imaging sequence involved more complicated spacecraft slewing than usual.

Nevertheless, in spite of the difficulties, the flyby was flawless. The imaging sequence began some five minutes before closest approach, when views of a crescent Mathilde were obtained at a resolution of about 500 m per pixel. The highest-resolution data (160 m per pixel) were obtained at closest approach (1212 km/753 miles), when the phase angle was close to 90° and Mathilde was half illuminated. The imaging sequence continued for another 20 minutes as the spacecraft receded from the asteroid.

Mathilde's mass was determined by accurately tracking NEAR before and after the encounter. Except for a short interval during the close approach period, continuous tracking of the spacecraft was performed from one week before, to almost one week after, the flyby. The tracking data led to a mass estimate for Mathilde of 1.033 ($\pm$0.044)$\times 10^{17}$ kg – a bit more than a millionth of the mass of the Moon. Coupled with a volume estimate from the imaging team, it yielded a surprising bulk density for Mathilde of 1.3 g/cm³, only slightly greater than the density of water.

On July 3, 1997, NEAR Shoemaker's large bipropellant rocket engine was fired for the first time. The seven-minute burn appeared to go quite well, imparting a ΔV of about 269 m/s. This substantial maneuver, which was followed by three smaller burns using the hydrazine thrusters, was to set up the gravity-assist flyby of Earth on January 23, 1998. The flyby was arranged so that NEAR and Eros would automatically approach each other travelling at similar speeds in the same direction. In addition to changing NEAR Shoemaker's trajectory, the Earth flyby provided an excellent opportunity for its instruments to make essential calibration measurements.

A botched rendezvous

In early December 1998, the mood in the NEAR project office was one of quiet confidence and anticipation of a successful orbit insertion at Eros on January 10, 1999. To carry out the rendezvous with Eros, a conservative plan calling for five maneuvers over a three-week period would begin with a

15-minute firing of NEAR Shoemaker's large engine. Although this burn would be about twice as long as the previous one with this engine, a smooth operation was expected.

NEAR Shoemaker's first rendezvous maneuver began on schedule at 5:00 p.m. EST (Eastern Standard Time) on December 20. Doppler signals confirming the start of the "settling burn" by the small thrusters began to arrive at the JPL navigation center at about 5:21 p.m. (One-way light time between NEAR Shoemaker and Earth was 20.7 minutes.) The settling burn proceeded as expected, and it appeared that the next step, firing the bipropellant engine, had begun on time. At this point, the operation started to unravel rather quickly.

Although sparse and inconclusive, the initial Doppler measurements indicated that the bipropellant burn was not normal. Worse yet, all contact with the spacecraft was lost 37 seconds after ignition. Another two hours went by before the navigation team was able to obtain a complete set of Doppler data, which were crucial to determining just what had gone wrong. These measurements showed that the bipropellant engine had shut down after less than one second. More ominously, large Doppler excursions had occurred just before the signal was lost.

A spacecraft emergency was declared, and the NEAR team worked throughout the night trying to restore communications. By the following morning, with NEAR Shoemaker still silent, the outlook was grim. The chief hope was that the transmitter had turned off autonomously due to a power shortage on the spacecraft. If this had happened, the transmitter would come back on-line automatically in 24 hours. Of course, this was only one of several theories under investigation. By Monday afternoon (December 21), many people feared that NEAR Shoemaker was gone for ever.

Then suddenly, at 8:01 p.m. on December 21, like a miracle, transmissions from NEAR Shoemaker were received at the Canberra station of the Deep Space Network. The spacecraft was still alive! After 27 hours of gloom and doom, joy had returned to the NEAR control center.

By early morning on December 22, enough engineering data had been downloaded from NEAR to permit an assessment of the spacecraft's status. There was good news.

•   There was no damage to the spacecraft or the propulsion system.

•   The fault-protection software had identified the problem and placed NEAR Shoemaker into a safe mode. The spacecraft's transmitter had turned off to save power and switched back on 24 hours later.
•   The cause of the engine abort was identified as a spike in lateral acceleration that exceeded limits set in the on-board software. It would be relatively easy to modify the software to eliminate this problem for future maneuvers.

And there was bad news.

•   The recovery was seriously deficient. Following the engine shut-down, the spacecraft experienced anomalous attitude fluctuations (spinning, tumbling), and the battery discharged to dangerously low levels.
•   The spacecraft expended roughly 30 kg of hydrazine fuel while stabilizing its attitude.
•   There was some contamination on the camera optics from the hydrazine fuel.

Naturally, the flawed recovery was a major concern. Another misfortune of this magnitude would effectively end the NEAR mission. Therefore, the project instituted a number of new operational and contingency testing procedures that would prevent a recurrence of this problem.

The loss of so much hydrazine was totally unexpected. Although a rendezvous with Eros was still achievable, the depletion of fuel precluded any possibility of a quick return. Had another 15 kg been squandered, NEAR would have lost all hope of ever attaining a rendezvous with Eros. Fortunately, the possibility of adversity had been built into the mission design. There were generous fuel supplies and a variety of contingency options. More than any other factor, the resilient mission design was responsible for giving NEAR a second chance. However, at that moment, there was no time for worrying about which contingency option would be used. Ready or not, a flyby of Eros would take place on December 23.

With less than 24 hours to prepare, engineers and scientists worked throughout the night to update NEAR Shoemaker's observing sequence. Unfortunately, the aborted rendezvous burn and ensuing attitude maneuvers had pushed the spacecraft far off its intended course. Instead of approaching to within 1000 km as originally planned, NEAR Shoemaker's minimum distance from Eros was 3827 km (2378 miles). This meant that the smallest detail resolved by NEAR Shoemaker's camera was

Figure 2.6. Revised trajectory profile for the NEAR mission. This modified version of Figure 2.1 takes into account the December 1998 main engine misfire and subsequent flyby of the asteroid. Fortunately, the main engine was fired correctly shortly thereafter, but then it took NEAR Shoemaker 13 months to finally catch up to Eros in February 2000.

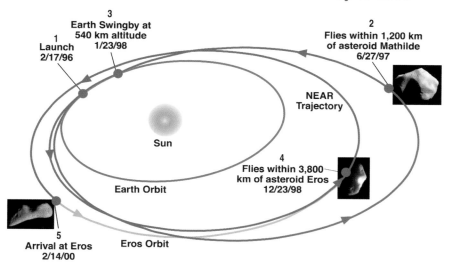

about 400 m (a quarter of a mile) across. Nevertheless, humankind's first close-up encounter with a near Earth asteroid yielded 222 images of Eros as well as supporting spectral observations.

While scientists were savoring the pictures of Eros, other members of the NEAR team (especially mission design, navigation, and operations personnel) were focused on formulating a plan to get NEAR Shoemaker back on track for a rendezvous with Eros. Existing contingency plans included options for a second attempt later in January 1999, in July 1999, or sometime between February and May 2000. Because of NEAR Shoemaker's unscheduled fuel dump, the January 1999 option was no longer possible, and the July 1999 option was marginal. The NEAR mission planners quickly settled on a strategy that would achieve a rendezvous in mid February 2000. The new target date was February 14, 2000 – Valentine's Day.

NEAR remained fairly close to Eros throughout 1999. Its movement away from Eros was reversed by a large propulsive maneuver on January 3, 1999, accomplished with the bipropellant engine, which this time performed flawlessly. There was a collective sigh of relief. All future thruster burns were to be significantly smaller, with net velocity changes of about 35 m/s compared to the 932 m/s for the January 3 maneuver. Furthermore, it would not be necessary to use the bipropellant engine to carry out these smaller maneuvers. A clean-up maneuver on January 20 and a mid-course correction on August 12 targeted NEAR Shoemaker to the vicinity of Eros.

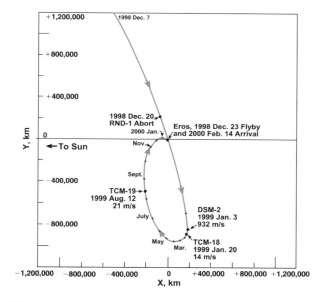

Figure 2.7. Trajectory plot of NEAR Shoemaker's December 1998 flyby and subsequent "U-turn" to catch up to Eros again in February 2000. The frame of reference is that of an observer orbiting with Eros around the Sun but positioned high above the asteroid. From this perspective, the spacecraft's aborted main engine burn on December 20, 1998 (RND-1 Abort) resulted in a close flyby and then the spacecraft continuing on past the asteroid for another 800 000 km (500 000 miles). Then, after the successful main engine firing on January 3, 1999 (DSM-2), followed by two smaller trajectory correction maneuvers (TCM-18 and TCM-19), NEAR Shoemaker appeared from this perspective to stop, turn around, and head back towards Eros. The spacecraft was eventually captured into Eros orbit on February 14, 2000.

Table 2.1. *Planetary Orbiters: Historic firsts*

1957 October 4	Earth	Sputnik 1 (USSR)
1966 April 3	Moon	Luna 10 (USSR)
1971 November 14	Mars	Mariner 9 (USA)
1975 October 22	Venus	Venera 9 (USSR)
1995 December 7	Jupiter	Galileo (USA)
2000 February 14	Eros	NEAR Shoemaker (USA)

Valentine's Day 2000

On its final approach, NEAR was to pass between the Sun and Eros at a distance of approximately 200 km from the asteroid. During this pass on February 13–14, the relative velocity between NEAR and Eros was only 10 m/s, and the phase angle of Eros as viewed from the spacecraft varied from 50° to about 1° at the end of the flyby. These special conditions allowed NEAR Shoemaker's infrared spectrometer to collect more than 5600 reflectance spectra of Eros at spatial resolutions between 2.5 and 12 km per spectrum.

Shortly after the "low-phase flyby," NEAR reached the desired orbit plane around Eros, which was perpendicular to the Sun's direction. It successfully executed the orbit insertion maneuver (a velocity change, ΔV, of about 10 m/s) at 10:47 a.m. EST on February 14, 2000. This maneuver placed NEAR into a 321 × 366 km elliptical orbit around Eros. Because there was still some uncertainty over Eros's mass, the orbit was more circular than originally planned.

The historic significance of NEAR's placement into an orbit around Eros should not be overlooked. This

was the first time that a spacecraft had orbited a small body. The accomplishment also made NEAR the latest member of a very select club of planetary orbiters (see Table 2.1).

Orbital operations at Eros

A series of small orbit correction maneuvers gradually brought NEAR Shoemaker closer to Eros until it reached its nominal mission orbit of 50 × 50 km on April 30. As the spacecraft descended through these early orbits, physical parameters of Eros, such as its mass, moment of inertia, gravity field, and rotation pole position, were determined with increasing precision.

When NEAR Shoemaker arrived in February 2000, Eros's north pole was oriented towards the Sun and its southern hemisphere was dark. About five months later, as a result of its motion around the Sun, Eros's rotation axis was perpendicular to the Sun–Eros line. NEAR Shoemaker's orbital plane was controlled so that it was always within 30° of a plane normal to the Sun–Eros line. In this configuration, NEAR Shoemaker's fixed solar panels are oriented towards the Sun. The science instruments are pointed at Eros's surface by slowly rolling the spacecraft as it orbits Eros. By February 2001, Eros south pole was directed at the Sun, and the northern hemisphere was in darkness.

The timeline for NEAR Shoemaker's orbital operations at Eros is summarized in Table 2.2. During the initial high-orbit phase, NEAR Shoemaker obtained a

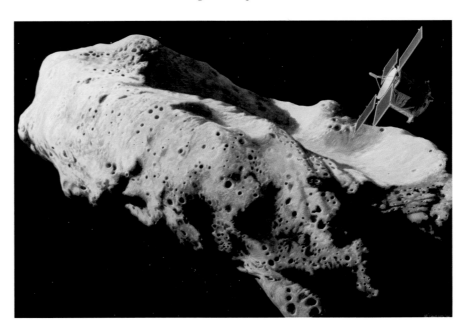

Figure 2.8. Artist's conception of the NEAR Shoemaker spacecraft in orbit around 433 Eros (painting by Pat Rawlings, 1995).

Table 2.2. *Operations timeline at Eros, 2000–2001*

February 14, 2000	Eros orbit insertion
February 14–April 30	High-orbit phase (orbit radius 50–365 km)
May 1–August 26	Low-orbit phase (orbit radius 35–50 km)
August 26–December 13	High-orbit phase* (orbit radius 50–200 km)
December 13–January 24	Low-orbit phase (orbit radius 35 km)
January 24–28	Low-altitude operations (minimum altitude 2.74 km)
January 28–February 11	Landing preparations (orbit radius 35 km)
February 12, 2001	Landing on Eros
February 13–28	Post-landing surface operations

Note:

*Except for a 5-km-altitude flyby on October 26.

considerable number of global images of Eros's northern hemisphere at resolutions of about 25 m per pixel. Later, when the spacecraft reached its 50 × 50 km orbit, NEAR Shoemaker's camera mapped the surface at scales of 5–10 m. Because NEAR Shoemaker's nominal orbit plane was close to the terminator plane (the plane dividing the dayside from the nightside), most of the images were taken at phase angles near 90°.

It was necessary to get closer to Eros not only to

obtain higher-resolution images but also because the X-ray/gamma-ray spectrometer (XGRS) required long observation periods in orbits with radii of 50 km or less. Only the lowest orbits would provide sufficient sensitivity and resolution for the XRGS instrument to measure and map the surface composition of Eros. However, the evolution of low-altitude orbits around Eros is strongly influenced by its irregular gravity field. Orbits exist that are quite unstable, and safe operation in these low-altitude orbits required close coordination between the science, mission design, navigation, and mission operations teams.

During the first low-orbit phase from May 1 to August 26, 2000, NEAR Shoemaker spent virtually all of its time in a 50 × 50 km orbit. The only exception was a brief ten-day interval in July when it operated in a 35 × 35 km orbit. Because the first operation in the 35 × 35 km orbit did not encounter any serious problems, it was decided to go directly to this orbit during the second low-orbit phase. This decision was significant because it allowed the XGRS instrument to operate for about two months in an orbit that regularly passed by Eros at altitudes under 20 km.

Of course, it was no surprise that the imaging team wanted to get even closer to Eros. To satisfy their desire for very high-resolution images, a 5-km-altitude pass, over one of the ends of Eros, was scheduled for October 26. This pass was executed

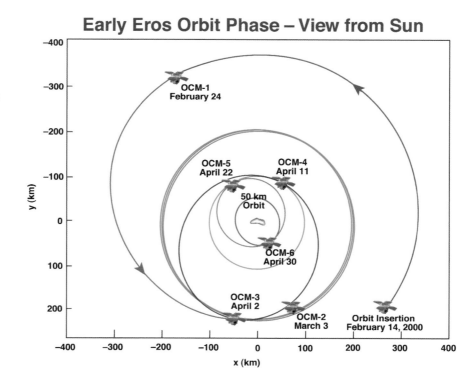

Figure 2.9. Trajectory diagram of the NEAR Shoemaker spacecraft's initial orbital operations at Eros, viewed from the perspective of an observer situated between the asteroid and the Sun. The initial orbit insertion and orbit correction maneuvers (OCM-1, OCM-2) placed the spacecraft in a 200-km (125-mile) circular orbit by early March 2000. In mid April the spacecraft was lowered to a 100-km orbit, and in late April the orbit was lowered to 50 km.

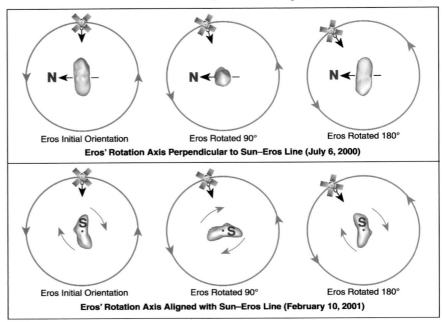

Figure 2.10. NEAR Shoemaker's orbital geometry at Eros in July 2000 and February 2001. As in Figure 2.9, the frame of reference is that of an observer situated between the asteroid and the Sun. The spacecraft's solar panels remained fixed relative to the Sun throughout the mission, since it required solar power to operate. However, because Eros's axis is tilted by nearly 90°, the "seasons" on the asteroid changed dramatically over the course of the mission. In July 2000 it was southern spring and Eros's equator faced the Sun. In February 2001 the season had advanced to southern summer and Eros's south pole faced the Sun. These seasonal changes resulted in large variations in lighting and viewing geometry over the course of the mission (see also Chapter 4).

Orbit size: 50 × 50 km; orbital period: 1.2 days

Eros Initial Orientation — Eros Rotated 90° — Eros Rotated 180°

Eros' Rotation Axis Perpendicular to Sun–Eros Line (July 6, 2000)

Eros Initial Orientation — Eros Rotated 90° — Eros Rotated 180°

Eros' Rotation Axis Aligned with Sun–Eros Line (February 10, 2001)

without any problems, and produced images with a resolution of better than 1 m. The success of the October 26 low-altitude flyover led to a second set of low-altitude passes in late January 2001 that culminated with a 2.74-km-altitude pass on January 28. The January 28 images revealed surface details at resolutions under half a meter.

The expenditure of fuel during NEAR Shoemaker's orbital operations at Eros was quite low. A total of 23 orbit-correction maneuvers were carried out through the end of January amounting to a total ΔV of only 17.2 m/s. The remaining ΔV capability was estimated to be 36.1 ± 6.8 m/s.

Landing and surface operations

One of the lingering questions that existed even before the NEAR launch was what should be done with the spacecraft when its primary mission was completed. Should it just be abandoned in its orbit around Eros? Alternatively, could a scientifically useful extended mission be identified? One idea that was put forward was to place the spacecraft in a stable heliocentric orbit where it could detect gamma-ray bursts with its gamma-ray spectrometer. This idea was certainly of some value scientifically, but it had nothing to do with improving an understanding of Eros. Furthermore, it would have been a very dull way to end a year of exciting discoveries in orbit around Eros.

A more adventurous proposal suggested that NEAR Shoemaker should slowly descend to Eros's surface and attempt a landing. During its descent, the spacecraft would keep its high-gain antenna pointed at Earth to transmit images and other science data as fast as possible. Although the landing idea would definitely attract considerable media attention, several key members of the NEAR team were less than enthusiastic. They were concerned that a landing attempt would be perceived as a high-risk stunt with marginal scientific value. They were also very worried that a failure would tarnish the favorable impression of NEAR's earlier successes.

On the other hand, supporters of the landing option argued that this was too good an opportunity to pass up. If everything went according to plan, images of Eros's surface with resolutions 10–20 times better than anything obtained earlier would be acquired. Because the images would be returned during the descent, success would not depend on the spacecraft surviving the landing impact.

After listening to all of the arguments, pro and con, NASA decided in favor of a "controlled descent" to Eros's surface. The primary goal of the controlled descent was to obtain high-resolution images. A secondary goal was to achieve a "soft" landing (that is, with an impact velocity less than 3 m/s). There was also a small hope that the spacecraft would survive the landing and transmit a signal from the surface.

The controlled descent to Eros's surface was scheduled for February 12, 2001, just two days before the formal end of the mission, when funds for mission operations were due to be cut off. This plan would allow a couple of days to contact the spacecraft after it had set down on Eros. Needless to say, there were very few people who dared to dream about this possibility. After all, NEAR Shoemaker was not designed to land, and it was quite likely that the spacecraft would end up on the surface with its solar panels and antennas facing down into the dirt.

The ΔV required for the landing maneuvers was estimated to be about 24 m/s, but this was comfortably within NEAR's minimum ΔV capability of 29.3 m/s. Because of power restrictions, only three instruments would be on during the descent, the imager, the laser rangefinder, and the magnetometer. To minimize the risk of losing communications during the descent, the timing of the landing phase was chosen to allow simultaneous coverage from two 70-m antennas of the Deep Space Network. On February 12, Eros was 2.11 AU from Earth, which corresponded to a one-way delay time of approximately 17.5 minutes for NEAR Shoemaker's transmissions to reach Earth.

In preparation for the descent phase, NEAR Shoemaker was placed into a 35 × 35 km orbit on January 28. A de-orbit maneuver at about 10:30 a.m. EST on February 12 began the descent. Approximately three hours and 45 minutes later, the first of four braking maneuvers was initiated. This maneuver occurred at an altitude of roughly 5 km, and slowed NEAR Shoemaker's rate of descent by about 6 m/s. After three more braking maneuvers, NEAR

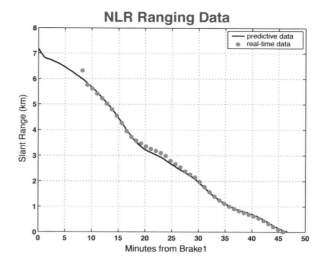

Figure 2.11. Data from NEAR Shoemaker's final descent to the surface of Eros, on February 12, 2001. The solid line is the descent profile predicted before the landing. The red dots represent the actual height measurements made by the NEAR laser rangefinder instrument during the descent. Everything went according to plan! "Slant range" is the distance to the surface along the spacecraft's current path; at the start the slant range of 7 km corresponded to an altitude of about 5 km.

Shoemaker touched down on the surface of Eros at 3:02 p.m. EST. On the way down, it returned 69 images of Eros, the last one snapped just 125 m above the surface.

Measurements from the laser rangefinder indicated that NEAR Shoemaker's braking maneuvers were very close to nominal. The nearly perfect performance resulted in an impact velocity of only 1.7 m/s. This was less than the 2.4 m/s impact velocity for Viking 1 on Mars, and could be the lowest ever. In any case,

Figure 2.12. Computer simulation of the NEAR Shoemaker spacecraft at rest on the surface of Eros. The simulation used the final descent telemetry information and the fact that the spacecraft remained in communication after landing to generate a best guess of the final configuration after landing. The tips of two of the solar panels and one corner of the spacecraft (and the NIS instrument) were probably gently buried in the regolith of the asteroid by the 1–2 m/s (2–4 mph) "impact." The final images suggest that the spacecraft landed on a "ponded" deposit of smooth and probably fine grained materials (see Chapters 4 and 7).

NEAR Shoemaker's touch-down on Eros was definitely a soft landing and not the "controlled crash" that some had predicted.

Not only had NEAR settled on Eros's surface very softly, it was still operating and sending a signal back to Earth! The spacecraft had made a gentle, picture-perfect three-point landing on the tips of two solar panels and the bottom edge of the main body. When it was determined that the gamma-ray spectrometer was still working, and could collect valuable data from an ideal vantage point about 10 cm from the surface, NEAR was granted a 14-day mission extension.

On February 28, commands were sent to place the spacecraft into a deep sleep. However, there is some possibility that NEAR Shoemaker may be heard from again. NASA may try to contact it one more time in September 2002 when NEAR Shoemaker is back in sunlight and Eros only 0.64 AU from Earth.

During its five-year mission, NEAR racked up an impressive list of "firsts", including:

- First spacecraft powered by solar cells to operate beyond Mars's orbit.

- First encounter with a C-class asteroid (June 27, 1997).
- First encounter with a near-Earth asteroid (December 23, 1998).
- First spacecraft to orbit a small body (February 14, 2000).
- First spacecraft to land on a small body (February 12, 2001).

The spectacular success of the NEAR mission was the product of many individual and collective contributions. It was also the result of a highly motivated team that succeeded in spite of considerable adversity – a team that was willing to accept risks in order to achieve a prominent place in the history of space exploration (see Table 2.3).

Table 2.3. *Planetary Landers: Historic firsts*

1966 February 3	Moon	Luna 9 (USSR)
1970 December 15	Venus	Venera 7 (USSR)
1971 December 2	Mars	Mars 3 (USSR)
2001 February 12	Eros	NEAR Shoemaker (USA)

3 From Launch to Rendezvous

Scott Murchie

Applied Physics Laboratory, Johns Hopkins University (NEAR Science Team member)

On February 17, 1996, the Delta 7925 rocket carrying NEAR Shoemaker leaped from its launch pad into a clear blue sky. In a minute all that remained on the ground was a smoke trail arching upwards. For scientists and engineers involved in NEAR Shoemaker's instruments, though, the launch was not the beginning of the mission and the years waiting for NEAR Shoemaker to reach Eros would be anything but quiet. Both before launch and during the cruise, there were thousands of preparatory tasks to be completed in order to fulfill the scientific objectives of the mission. NEAR was to orbit a city-sized asteroid and, while a hundred million miles from Earth, make detailed, accurate measurements of the asteroid's surface features, shape, and composition. The main tools for doing this work were the instruments on board the spacecraft.

My job on the project was as instrument scientist for NEAR's camera, the multispectral imager (MSI). MSI was effectively a digital camera with a telephoto lens. The body of the camera was a telescope 30 cm (one foot) long and images were recorded by a charge-coupled device (CCD) detector, similar to the ones found in digital cameras and camcorders. MSI had a filter wheel, allowing each individual black-and-white picture to be taken through a different color filter so that composite color images could be reconstructed later. The camera had a small computer attached, called a digital processing unit or DPU, which ran the operating software.

Humans are accustomed to viewing the world with two-color stereo imagers – our eyes – and our brains interpret this kind of information instinctively. We can glance at a surface and almost without thinking distinguish light materials from dark materials, red things from blue, rough surfaces from smooth ones, and rounded objects from angular shapes. For an instrument on a spacecraft, we have to simulate what our brains do using algorithms coded on computers.

Making NEAR Shoemaker's camera work effectively required a detailed understanding of exactly how it translated a scene into electronic signals and a practiced finesse in using the sophisticated software. We gained an understanding of how the camera performed by running a long series of controlled calibration tests. Many of these were done on ground in a laboratory during the summer of 1995 – almost a year before NEAR Shoemaker was launched. Others were done after launch, to track any changes in the camera and to do tests under actual operating conditions – the cold, weightless vacuum of space. A practiced finesse in running the camera could only be attained by practice, practice, practice.

While NEAR Shoemaker cruised through space towards Eros, dozens of calibrations had to be performed. The plan for capturing thousands of images of every kind – color images, stereo images, images taken with lighting from every possible direction – had to be established ahead of time, even though no-one had ever actually *seen* Eros or knew which were the key spots to image. And just as if the mission to Eros were a school play, there had to be numerous rehearsals. Everything was prepared so that once we were in orbit around Eros the busy year of picture taking would run like a well-oiled machine.

Cruising in space: February 1996–June 1997

The first test came four days after launch, on February 21. The objective: simply to take a set of pictures of Earth's Moon, looking back over our shoulders a million and a half miles. It was easier said than done. Consider all the muscles and bones a human uses to turn around, look at something, focus his or her eyes on it, and watch it move across the sky. On NEAR

Shoemaker, as on any other spacecraft, every individual action required a command that is written on the ground. The list of commands – the script – had to have everything correct and in the right order. The spacecraft had been off the ground for less than a week and was being used for the first time as a science tool, so even a simple operation like taking pictures of the Moon was bound to be a learning experience and good practice for the many thousands of times the spacecraft would be turned to take pictures while in orbit around Eros.

Taking MSI images of the Moon was a valuable calibration too. Unlike Earth, with its dynamic atmosphere and weather systems, the Moon always looks the same. Samples of lunar soil tested in laboratories tell us the exact colors of the different parts of the Moon that have been visited by astronauts. By taking pictures of these areas, or areas just like them, we would have a good measure of how MSI "sees" colors. It hardly mattered that on February 21, 1996, NEAR Shoemaker was nearly 3 million km (1.8 million miles) from the Moon – eight times farther away than the distance between Earth and the Moon – and that the lunar orb was a mere 13 pixels across as seen by MSI.

Development of the script for NEAR Shoemaker's first picture-taking excursion began over four months before launch, squeezed in among exhaustive preparations for actually launching the spacecraft. The procedure foreshadowed what the first part of the mission in space would be like: the scientists developed an observation plan, Mission Operations engineers wrote and edited the script of commands, the script was reviewed by Ed Hawkins (the lead engineer who built MSI and who knew its every technical detail) and by me, and finally the script was tested on the on-ground simulator of the spacecraft's computer systems (the "brassboard"). When the first pictures came down on February 21, everyone gathered in Mission Operations. The spacecraft lead engineer, Andy Santo, smiled when we all saw the Moon dead center, exactly where it was supposed to be, and then laughed and shook his head – "13 pixels! Man, you guys. You did all that work for *this*?" Joking aside, we all knew that the Moon's size did not matter. What mattered was that NEAR Shoemaker worked and obeyed its commands perfectly.

The next test came a little over a month later – just when the overworked staff of NEAR had launched the spacecraft, taken the Moon pictures, and were actually thinking about getting some rest. At that time, in March 1996, Comet Hyukatake was passing near

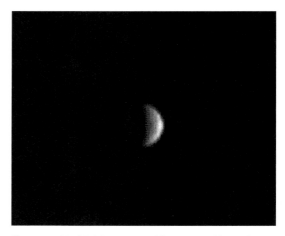

Figure 3.1. Earth's Moon, imaged from NEAR Shoemaker at a range of 2.9 million km (1.8 million miles), on February 21, 1996.

Earth, cutting across our planet's orbital plane. The comet's speed was so high that, night to night, it moved significantly across the sky. Joe Veverka of Cornell University, the head of the team of scientists responsible for calibrating MSI and using it to take pictures of Eros, put it this way: "Congratulations. With the Moon you've proved that you can take a picture of a stationary target. Now let's do something more like what we'll do at Eros – take pictures of a moving target. I want a mosaic of the comet, and I want it in color."

Two aspects of the Hyukatake drill were particularly valuable. Imaging the comet was an exercise in predicting the future location of a moving object, pointing MSI at it, and taking a mosaic of pictures. The Moon images had been just single frames, not a mosaic of the kind we would have to use to image Eros from close up. Less obvious was the practice in teamwork. This early in the mission, the people who ran NEAR and MSI were two disparate groups: engineers and mission designers at Johns Hopkins University's Applied Physics Laboratory (APL), who actually conducted the mission, and the outside scientists whose job it would be to analyze NEAR's returned data. Everyone had met at formal meetings and watched each others' viewgraph presentations. Some science team members had even participated in the on-ground testing of MSI prior to launch. But this would be the first exercise in functioning as a team. One thing that definitely was not a goal of the comet images was hard-core science. NEAR Shoemaker was so far from the comet that, at best, it would be a fuzzy streak in the images.

Even at this early time, it was clear who would be the primary players in running the mission. On the spacecraft side were Andy Santo, Ed Hawkins, Rob Gold (the overall manager for the science payload), and Tom Coughlin. Tom was the project manager who worked mostly behind the scenes, making sure that people did their jobs and that bills got paid. On the science side were Joe Veverka, Mark Robinson (then at the US Geological Survey and later at Northwestern University), Jim Bell (then at NASA Ames Research Center and later working with Joe at Cornell University), Peter Thomas and Ann Harch, both also at Cornell, and myself. Everyone had their science role. Mine and Jim Bell's were to study color in the images. Mark's and Peter's were to use the images to study shape and landforms, and Joe's was to put all of our findings into the big picture. Joe's decades of experience on previous space missions, including Voyager and Galileo, were marked by a stunning record of publications and had earned him a unique reputation as someone who could motivate people using both the carrot and the stick. Ann's job was – well – everything. She had designed the only previous imaging sequences of asteroids, the Galileo spacecraft's fleeting glimpses of main-belt asteroids Gaspra and Ida as it flew by them at high speed en route to Jupiter. Starting with Hyukatake, she honed her skills on NEAR, taking overall responsibility for planning what pictures would be taken, and when.

As with the Moon images, the command scripts were written and reviewed, re-reviewed, and tested. When the night of the Hyukatake observations came, the weather at APL was cold and clear. On the ride into work for the 1.00 a.m. operation, Hyukatake was spread across the Maryland sky. It was nice to see it with real eyes and then again, hours later, with electronic eyes from millions of miles away. When the images were downlinked to Earth, the fuzzy little streak was dead center in the middle of the mosaic, surrounded by two-thirds of a million pixels of black space. Within hours, Jim Bell produced an annotated image in which Hyukatake was surrounded by faint stars.

So far, so good.

Two and a half months after launch, at the end of April, 1996, inflight calibration started in earnest. As a check on the calibration of MSI's response to color using the Moon images, and to track any changes, a series of color imaging sessions of the bright star Canopus was begun. This series was to last until Eros encounter. We also took images of large clusters of stars. MSI's design parameters and on-ground images

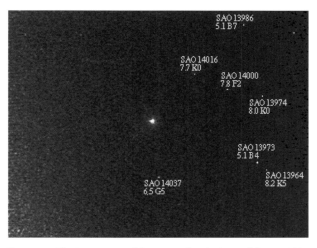

Figure 3.2. The inner coma of Comet Hyakutake imaged from NEAR Shoemaker on March 24, 1996, while the spacecraft was 16.4 million km (10.3 million miles) from the comet. Stars in the background are named.

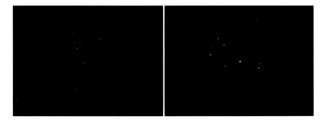

Figure 3.3. (Left) The Beehive Cluster (Praesepe), imaged from NEAR Shoemaker on May 2, 1996. (Right) The Pleiades star cluster imaged from NEAR Shoemaker on January 15, 1997.

of small geometric patterns told us, to within a fraction of a percent, how much magnification was provided by the telescope. But almost unbelievably, it can be determined more accurately in space (to better than one part in a thousand) by imaging star clusters, because the apparent distances between their stars as viewed from Earth have been measured with extraordinary precision. As part of the first calibrations on April 29 and May 2, Praesepe (the Beehive Cluster) was imaged. The Pleiades were imaged on January 15, 1997.

During those early calibrations in 1996, we also made a decision that seemed right at the time but would later be nearly devastating. Before launch, mission director Bob Farquhar and mission designer Dave Dunham discovered that NEAR Shoemaker's trajectory would bring it close to an asteroid named 253 Mathilde. Eros, like Gaspra and Ida, is one of the S-type asteroids common in the inner solar system, made of the same kinds of rock and metal as Venus, Earth, and Mars. Mathilde, on the other hand, is a

C-type made of carbon-rich rocks that are as dark as soot. It has more in common with comets than with Earth's rock types. This would be the first chance ever to see a C-type asteroid up close. And NEAR Shoemaker could alter its trajectory enough to pass close by Mathilde and get a great view, at the cost of a small amount of fuel.

The only problem was that imaging Mathilde was at the ragged edge of what NEAR Shoemaker, and MSI, were designed to do. Sunlight at Mathilde's greater distance from the Sun is less than half as intense as at Eros, and the rock Mathilde is made of is eight times darker, so to take a good picture of Mathilde you have to expose longer than for Eros. MSI was designed to cover Eros with imaging from close orbit, so the camera's field of view was wide-angle by the standard of planetary cameras (about 3°). What is more, NEAR Shoemaker was built with the capability to turn only very slowly – at the fraction of a degree per second it needed to image mosaics of Eros while orbiting at a velocity measured in meters per second. But NEAR Shoemaker would fly past Mathilde at 10 km/s. There was nowhere near enough fuel on the spacecraft to reduce the flyby velocity. Just to complicate things further, we would only know approximately where Mathilde was. To be sure of getting an image, we would have to image the whole volume of space in which the asteroid could be – a region called the uncertainty ellipsoid.

To see small details NEAR Shoemaker had to get close to Mathilde. To be close to Mathilde going so quickly, and to take unsmeared images, meant doing three things: turning the spacecraft fast enough to track the asteroid, being far enough away to be able to pan repeatedly across the uncertainty ellipsoid, and taking very, very short exposures. Gene Heyler, one of the chief architects of NEAR Shoemaker's guidance and control system, determined that 1200 km was the right flyby distance for panning the camera and being able to cover the ellipsoid repeatedly. However, the exposures would have to be so short that the images would be disappointingly dim. If only there were a way to make the camera more sensitive.

In fact, there was a way. MSI was launched with a cover designed to protect its optics from any contamination during the firing of NEAR Shoemaker's main engine. The cover had a small, clear aperture with a quartz window that let through about one-fifth of the incident light. The plan had always been to leave that cover on until we were almost at Eros and the firings of NEAR Shoemaker's big main engine had been

completed. The temptation to blow the cover early and get better images of Mathilde was irresistible. At the meeting where this was decided, just after launch, Rob Gold and Joe Veverka got into an intense discussion about what to do. Rob maintained that the cover should stay on until Eros, and the Mathilde images allowed to suffer. NEAR was a mission to Eros. Joe countered that this was a mission of exploration, and we could explore the unknown at Mathilde, so it was justifiable to blow the cover now. Joe's arguments carried the day. Amidst the calibrations of May 1996, we blew it – literally and figuratively. We would get great images of Mathilde but unbeknownst to us, we would have to pay the price three years later.

Waltzing with Mathilde

Preparations for flying by Mathilde went on for a full year. During that time, the mission was transformed. When the first test was done in June 1996, pointing just off the Sun to take incoming navigation images, the strain of launch and early operations was beginning to show on everyone. This was especially true of the overworked staff at Mission Operations. When NEAR was conceived and the staffing assembled, it was strictly a mission to Eros. The flyby of Mathilde had not been dreamed of and, to keep the mission simple, a mantra had been "no cruise science." Now the skeleton Mission Operations staff, whose numbers had been planned on the basis of a relatively uneventful three-year cruise, was being asked to prepare for a new encounter in a year and at the limits of NEAR Shoemaker's capability. They had not yet had the time to set up a ground system to automate the generation of command scripts, so everything still had to be done by hand. The work load was so overwhelming that before long only one original member of the Operations team remained, Bob Nelson. New staff were hired, including Charles Kowal, discoverer of the strange, asteroid-like object Chiron, which orbits the Sun way out between Saturn and Uranus.

Finally, in July 1996, something gave. A command to point the spacecraft's solar panels too far off the Sun was erroneously sent. There was insufficient review of the commands and NEAR Shoemaker went into safe mode. The engineers who built and programmed NEAR Shoemaker had given it a detailed set of failsafe rules for protecting itself – "autonomy rules" – largely the brainchild of software lead Sue Lee. NEAR Shoemaker knew its limits. Most, like how

Figure 3.4. A schematic representation of NEAR Shoemaker's encounter with the main-belt C-type asteroid 253 Mathilde on June 27, 1997. The spacecraft flew past the asteroid at a range of 1200 km (750 miles) with a speed of about 10 km/s (22,500 mph). The entire close encounter sequence was over in about 25 minutes.

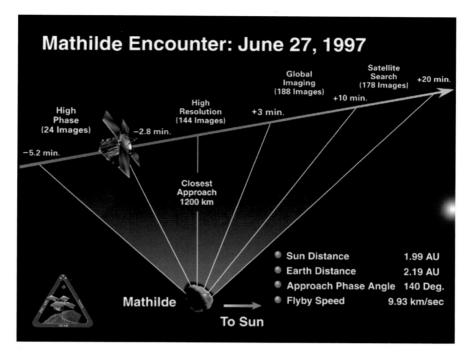

far off the Sun it could point the solar panels, could be set from the ground and if it was told to exceed those limits it would refuse and instead go into a standby mode waiting for correct instructions. This does not endanger the spacecraft. In fact it protects it, but the snag shook up everyone.

Quietly, a new lead for Mission Operations was recruited. When Mark Holdridge took over in early 1997, there was a sea change in the way of doing business. Soon too came new faces who later formed the backbone of NEAR operations. One of them was Karl Whittenburg. To those of us who worked on the inside of the mission, Karl *was* NEAR. Name almost any aspect of the spacecraft or its software, and Karl knew about it. In the unlikely event that someone mentioned some feature that Karl did not know intimately, by the next day he had made himself expert in it. Bob Nelson was the same. With Mark, Bob, and Karl in Mission Operations, the new mantra was "practice, practice, practice." In the months leading up to the Mathilde flyby, everything was practiced. In the two days prior to the flyby we would take images every six hours. I would process them and Bill Owen at JPL would locate Mathilde's position relative to the background stars. Using this approach, if in the real event we found ourselves off-course, a last-minute course correction could be made. To practice acquiring and processing the images, we did two rehearsals by hooking up the engineering model of the MSI to the brassboard and taking pictures of a wall clock, taking things just as

seriously as if NEAR Shoemaker were really just a day away from Mathilde.

Meanwhile, Ann Harch designed the imaging coverage of the uncertainty ellipsoid. She and Gene Heyler created a sequence that accounted for anything: if Mathilde were in the middle of the ellipse we would get ten views of it and if it were off in the corner we would still get at least eight. After close approach, when the lighting was best, there were color sequences in case Mathilde had compositional variations on its surface. From 10 to 20 minutes after closest approach, there were two searches for satellites, the first to catch a color close-up of any moonlet near to Mathilde and the second to look for little moons further out.

Five partial or full dress rehearsals were done, called "Shamtillys". This name was Gene Heyler's creation: "Sham" for fake and "tilly" for Mathilde. Each Shamtilly was rehearsed on-ground first, then in-flight. The first ones did not work perfectly but, by the fifth one, Gene's spacecraft attitude plots showed the spacecraft turning to the right place at the right time, and Bill Owen located stars in the practice pictures of black space to verify where the camera was pointed and to assure us that, if Mathilde had been there, we would have seen it.

The real thing started on June 25. Out-of-town team members flew into Baltimore. Project scientist Andy Cheng had arranged a temporary science center in a prefabricated building at the edge of the APL

Figure 3.5. The third image mosaic of Mathilde, taken from a range of 1800 km (1120 miles). The lower half of the mosaic was the first image played back.

campus and Doug Holland, who ran the on-ground data system, set up enough computers for everyone to have a monitor to work on when the images came down. On June 26, Bill Owen located Mathilde in the second of six sets of optical navigation (OpNav) images, right where it should be. NEAR was on course, and there was no need for any last-minute correction.

Then fate intervened and the benefits of all the practice paid off. A thunderstorm broke with marble-sized hail, causing a massive blackout around APL. The temporary science center was plunged into dark-ness. Mission Operations was using uninterruptable power, so Bill Owen, out in California, simply down-loaded raw images from Operations via the internet and continued tracking Mathilde with no disruption. When the power came back on, Bill's e-mails reported that everything was looking great – except that Mathilde was not nearly as bright as expected. On Thursday night, hours before closest approach, Mark Robinson, Ann Harch, Jim Bell, and I went to Mission Operations to see what was happening. Despite the storm, the encounter had been so well rehearsed that everything was unfolding exactly as it should have. Bob Farquhar, Mark Holdridge, and others were eating pizza, watching the monitors, and smiling.

On Friday morning, the encounter was over before the first pictures had reached the ground. The whole of the science from the Mathilde flyby sat on NEAR Shoemaker's solid-state recorder, to be played back twice after the flyby. The double playback allowed any portion of an image lost in transmission the first time to be received correctly a second time. The playback started partway through the images, near the ones taken just after closest approach, so that the best data would have reached the ground first in the case of a spacecraft upset later. All of the science team and many APL engineers and staff were gathered in the temporary science center anticipating what Mathilde would look like.

When the first image appeared on a monitor, a cheer went up and Ann Harch burst into joyful tears. There was the bottom half of the second closest view of Mathilde, filling the screen. Right away the murmur went around the room, "Where is the rest of the aster-oid?" Half of the part of the surface that should have been illuminated was dark, hidden by shadow filling giant impact craters. Mathilde had been pummeled; its surface was so full of large craters that, literally, no more could have fit. Within hours, we were asking, "Why is this asteroid still in existence at all?"

The remaining images underlined how Mathilde had sustained a beating and yet had survived. The first mosaic, taken while approaching the night side and expected to show a small crescent, instead showed a bizarre form, promptly named "the pterodactyl" by Mark Robinson. Impact craters had taken huge bites out of the asteroid. Mathilde had seemed unexpectedly dark in the OpNav images because most of it was not there! The next mosaic, taken at the time of closest approach, was more *shadow* than asteroid: most of this view of the surface was dominated by enormous craters whose interiors, shielded from sunlight, were black! The subsequent mosaics revealed the origin of the surreal, contorted form of Mathilde: at least five large craters, each comparable in size to Mathilde's radius, shape the asteroid's mountains and valleys. Even the asteroid's profile has been molded by the craters. But after detailed examination of all the mosaics by Clark Chapman and Bill Merline, team members from the Southwest Research Institute, not even the tiniest moonlet nor any leftover piece of debris could be found.

After the dance

Three words sum up Mathilde: big, black, and battered. Mathilde is by far the largest asteroid so far encoun-tered by a spacecraft. With an average diameter of

Figure 3.6 (Left) The first image mosaic of Mathilde, taken over the asteroid's night side. (Right) The second image mosaic of Mathilde, taken at the time of closest approach to the asteroid, from a range of 1212 km (750 miles). From this viewpoint, most of the surface lies shadowed in the interiors of enormous craters.

Figure 3.7. Asteroids Mathilde and Ida, shown at their correct relative brightnesses.

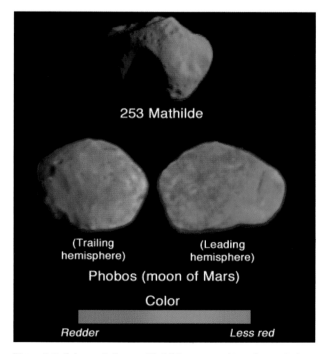

Figure 3.8. Color variation on Mathilde compared to color variation on the martian moon Phobos, imaged by the Russian spacecraft Phobos 2 in 1989.

52 km (32 miles) it dwarfs tiny Gaspra, and is as wide as the state of Rhode Island. One of its huge craters could easily swallow Washington, DC.

Mathilde is by far the darkest asteroid imaged by a spacecraft, darker than soot. Astronomers expected to find that Mathilde is plain and colorless. Their reasoning was that Mathilde's low reflectivity was probably due to a high content of carbon (which is also abundant in the primitive, so-called carbonaceous, meteorite class). Carbon and carbon compounds are gray and colorless at the wavelengths measured by MSI, and they overwhelm whatever rocks or minerals they are mixed with. And Mathilde *is* as colorless as it is black. In fact, Mathilde holds the dubious distinction of being the only rocky solar system body yet imaged in color without *having any* color. Gaspra and Ida have fairly obvious color variations associated with young craters. But all the computer enhancements of color pictures of Mathilde reveal only tiny color aberrations in NEAR Shoemaker's camera. After Mathilde, the next darkest asteroid-like body visited by a spacecraft is Mars's inner Moon, Phobos. In 1989 the Soviet spacecraft Phobos 2 made three close flybys of Phobos, each time taking color images. Compared with Mathilde, Phobos is nearly twice as bright and its colors are spectacular.

The big discovery about Mathilde from the NEAR imaging data was the degree of cratering. The density of smaller craters (number of craters per square kilometer) is high, comparable to that on the oldest parts of the Moon, Mars, and Mercury – all ancient worlds. But the density of Mathilde's largest craters exceeds that of any other known planet or asteroid. Mathilde is "geometrically saturated" with large craters. No more could fit because new craters would just obliterate the old ones. The asteroid's shape shows a hint of once having been roughly spheroidal, but giant pock-marks have cut huge chunks from its outline. Yet, despite the violence, there is no hint of debris. There are no color variations to suggest that a different rock type has been excavated from the depth, no obvious rings of debris surrounding craters of any size, and no moonlets.

How can such a battered body still exist and not have been blown to pieces? And where did the material excavated from the craters – the "ejecta" – go? NEAR's radio science experiment provided a clue. During the flyby, Mathilde's gravity caused a slight acceleration of the spacecraft toward the asteroid. After the flyby, Mathilde's gravity slowed NEAR Shoemaker down by a similarly tiny amount. This

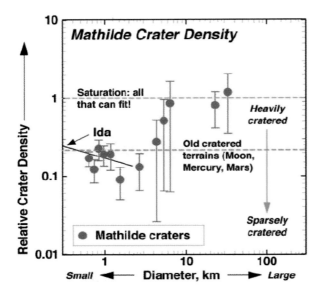

Figure 3.9. The relative densities of craters of different sizes on Mathilde (red dots), comparing the asteroid to Ida and to cratered surfaces on the Moon, Mercury, and Mars. On this graph by Clark Chapman and colleagues, the top green dashed line represents a cookie-cutter-like pattern of craters packed shoulder to shoulder, which is close to the actual situation with Mathilde's large craters.

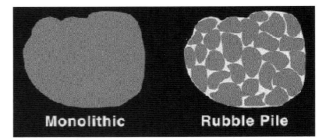

Figure 3.10. Mathilde's continued existence after its severe battering can be explained if the interior is a loose rubble pile of fractured rock, which could deform to absorb the shocks caused by impacts.

acceleration and deceleration caused the radio signals emitted by the spacecraft, as tracked on the ground, to change frequency, a phenomenon known as the Doppler effect. The ground receivers for planetary spacecraft in NASA's Deep Space Network (DSN) of antennas in California, Spain, and Australia can track Doppler shifts in frequency equivalent to the spacecraft accelerating by as little as a tenth of a millimeter per second (one foot per hour). By combining NEAR Shoemaker's distance from Mathilde, the MSI images, and the accelerations from radio tracking, Mathilde's mass was determined by Don Yeomans of JPL and his Radio Science team. Peter Thomas and colleagues at Cornell determined the asteroid's shape from images, providing an estimate of Mathilde's volume. Put mass together with volume and you have Mathilde's density – only 1.3 times the density of water. That is only half the density of even the lightest rock type that could reasonably be expected to make up its interior.

The inside of Mathilde must be a giant pile of loose rubble full of holes! Astronomers had long debated which asteroids might be solid chunks of rock and which might be so-called rubble piles. The low density of the asteroid provides almost unambiguous evidence that Mathilde is a rubble pile. And this provides a possible explanation of why Mathilde still exists: it probably withstood its pummeling because an interior only loosely held together absorbs the seismic energy from

impacts. The craters probably did not excavate rock but simply pushed in loose debris. In much the same way, a sandbag absorbs the blow from a sledgehammer that would shatter a solid rock. Shortly after the encounter, Joe Veverka made the connection between low density and Mathilde's resistance to disruption by the impacts that created the giant craters. His analogy was that southern forts in the USA were once built of porous, squishable palmetto logs because a cannon ball would simply push into the logs and embed where it hit, while a sturdy oak fort would be shattered.

Earth swingby: dress rehearsal for Eros

By January 1998, NEAR Shoemaker had completed two-thirds of its journey to Eros. The mission engineers, programmers, planners, and scientists had tested the performance of the spacecraft and of MSI during the cruise and the Mathilde flyby. But Mathilde was very different from Eros. Being so far from the Sun, the power from NEAR Shoemaker's solar arrays was limited and MSI had been the only instrument running. And NEAR Shoemaker's relative motion past Mathilde was very fast, much more so than it would be at Eros, allowing for less than an hour of data-taking. The coming swingby of Earth for the gravitational assist to Eros presented the first good opportunity for a dry run of operations like those that would take place at Eros, with all instruments on and taking coordinated measurements. Earth and the Moon were also the perfect targets: because they are well understood, we had a "cheat sheet" for how all the results should look.

Besides, in the case of MSI, there were one or more problems of unknown scope that could be addressed during the swingby. First, in the attempt to tease any sort of color variation out of Mathilde's surface, we found that scattered light in the camera was a real concern. In our images Mathilde had a blue halo around it, not because of anything real but because blue and green light was being scattered by MSI's optics much more than red or infrared light. We had not seen this in ground testing, mainly because of the characteristics of the chamber in which we had tested MSI before launch. That was a cooled vacuum chamber the size of a VW bus, in which the camera was mounted on a rotating table. Looking different ways, we could image different kinds of light sources to do an on-ground calibration of how the whole of MSI responded. But the inside of the chamber was not sufficiently black and dark to characterize scattered light inside the camera. We could never know for sure which faint patches of light in an image were due to the camera itself and which were due to light bouncing off the walls. During the flyby, Earth and Moon would be ideal test targets: nice and bright, crisp edges, seen at different sizes at different distances, and viewed against the perfectly black background of space.

The second problem was much harder to get a handle on. In a few of the images taken during the pans across Mathilde's uncertainty ellipse, there was a very faint "ghost image" of Mathilde a few degrees away from the real asteroid. Mark Robinson had found it first and we had brought it to Ed Hawkins' attention. Ed went back to the original optical designer of the camera to try to track down any possible reflections inside the optics. No source could be found. All we knew for sure was that there was something at least a little wrong.

The plan we came up with was to take lots of color images of Earth, starting just after the flyby when NEAR Shoemaker was still close to Earth and continuing for three days until a tiny Earth had receded into the distance. We were departing directly over Antarctica, so we would track the color of the snow and the extent of any halo around Earth's limb to better understand the scattered light. To look for ghost images, we planned a giant mosaic of overlapping images around the Moon in the shape of a plus sign. We expected to find the same ghost seen at Mathilde, but this time we would accurately locate it, track down its source, and see if any other ghosts were haunting the camera. The Earth swingby was also to be an end-to-end test of the instruments and our ability to deal with the data.

After lots of planning, Ann Harch created one of her usual brilliant plans. NEAR Shoemaker would pass low over Siberia, then zoom southwest and head out into interplanetary space over southern Africa. We

planned pictures and spectra from NEAR Shoemaker's near-infrared spectrometer (NIS) over all of the spots least likely to be obscured by clouds – Kazakhstan, Iran, Kuwait, and Saudi Arabia – plus a multi-instrument map using both MSI and NIS of South Africa and Namibia. Departing Earth, we would take a movie of the receding planet, breaking off only long enough to take the big lunar mosaic. The whole sequence would run from January 23 to January 26. As with Mathilde, there were endless on-ground reviews of command scripts, plots of simulated coverage, and finessing of the sequence.

The first playback of data on Saturday morning, January 23, was a great success – if you judge this kind of a test by what it teaches you about what still needs work. Data reaching the ground were passed to DSN and then to Mission Operations and finally to Doug Holland's ground data system. The link between Operations and the data system proved to be too slow to handle the data; the first playback showed up briefly in the data archive, then disappeared, and nothing further showed up as the data system ground to a halt. When the data were recovered from the DSN, we got our first good look at the images. Kazakhstan was clouded over, but we got a great view of Saudi Arabia, even seeing some human-made irrigation features in the desert. All of the NIS spectra looked beautiful – showing evidence of water vapor in Earth's atmosphere and water ice covering Antarctica. But the South Africa MSI color sequence was a shock. A few of the black-and-white frames looked OK, and the coordinated observations had executed perfectly. But the color images of South Africa, as well as the first few of Antarctica, had brilliant blue and red horizontal bars across the top and bottom. They were so intense it was hard to see Earth through them. The bars disappeared only gradually during the Earth movie, as Earth got smaller in the frame and Antarctica stopped dominating the scene. At first we had no idea what was going on.

Mark Robinson and I set up a speaker phone to call Joe Veverka, who was at home outside Ithaca, New York. "Hi, Joe, are you sitting down? You'd better." "OK, what's happening?" "We've got a little problem with the camera. Actually, the pictures of South Africa look like dirt." Silence.

On Monday morning Rob Gold convened everyone involved in designing the camera to get to the bottom of the horizontal bars. A new recruit at the lab, Allen Keeney, was given the responsibility of testing the flight spare camera using different illumination

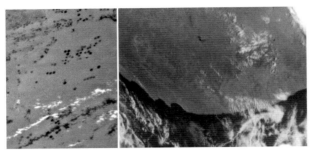

Figure 3.11. (Left) Image of the region southwest of the Saudi Arabian capital of Riyadh, taken from a range of 685 km (430 miles) on January 23, 1998, just after the spacecraft's closest approach to Earth. The dark circular features are irrigated fields, roughly 65 m (215 feet) in diameter. The white streaks running diagonally near the bottom are clouds. North is up, and the field of view is approximately 3 km across. (Right) Image of the south coast of South Africa acquired after closest approach at a range of 18 000 km (11 300 miles) as NEAR Shoemaker was receding from Earth. North is up and to the right and the field of view is approximately 720 km (450 miles) across.

patterns to try to reproduce conditions that led to the bars. At the same time, the Moon mosaic came down and provided the first good clue to the problem. The ghost was still there, smeared out into what looked like a short segment of a horizontal bar in most images. Allen's tests took a few months but eventually he managed to pinpoint what had happened to MSI. Light from just outside MSI's field of view, in one direction, was leaking into the light-sensitive electronics just after an image was exposed. In the pictures taken in color, the filter wheel was moving and that smeared out the light into a horizontal bar. The smeared ghost was worse when the light coming into the camera was brighter. It was overwhelming when MSI was looking at the bright white snow in Antarctica but barely detectable when the camera was directed at the dark surface of the Moon. Compared to Earth, Eros is dark and far from the Sun, so this problem would be only minimal. We would not be able to see the problem in black-and-white images, only when images through several filters were put together for color pictures.

Allen's diagnosis meant that there was an easy workaround. Because of the way MSI worked, we could take a long exposure of a particular scene then subtract a short exposure to eliminate the ghost image. The bad part was that this was a new, extra step in handling our pictures.

The Earth flyby gave the team the minimum practice it needed to be prepared for doing science at Eros and an awareness of MSI's "personality." Much of the data taken by the different instruments never made it past viewgraph presentations at science team meetings

Figure 3.12. A montage of images of Earth and the Moon acquired by NEAR on January 23, 1998 from a distance of approximately 400 000 km (250 000 miles). The Antarctic continent can be seen near the center of Earth's sunlit disk, and the reddish soils of Australia near the upper-right part of the disk. This composite view is not to scale; in reality the Moon is about ten times farther away and much darker than shown here.

or graphics in arcane papers on instrument calibration, but, nevertheless, they showed that NEAR Shoemaker worked.

Eros first light

The Earth flyby was the last holdout of the old way of doing business on NEAR, with long hand-done command scripts reviewed endlessly by committee. At Eros, we would need a simpler, more automated, way of commanding the spacecraft. Such a tool had been planned but, in the pandemonium of the early mission, it never got anywhere. So, it was decided to adapt for NEAR the tool that had been used on Mars Pathfinder, Galileo, and other previous planetary missions, an enormous piece of software called SEQGEN (pronounced "seek-jen"). This activity was led by Karl Whittenburg and Ann Harch. The science team was led through practices in using the totally user-unfriendly software over the summer. All other activities in early 1998 became secondary, except for planning the upcoming arrival at Eros beginning December 20.

The total switch to SEQGEN was supposed to happen in September. It did not. It materialized that the very first week of commands contained one that would have put the X-ray spectrometer into a dangerous state. Minutes before they were sent to the spacecraft, payload manager Rob Gold vetoed the whole week of commands. A whole new layer of review was put in place to ensure that no more episodes like that could happen.

Finally, November 5 rolled around. That was the day of the first, distant image of Eros, to be used for optical navigation. There was Eros, at last, a pinprick of light against the starry background from a range of 4 million km (2.5 million miles). Every week, then every day, another set of images was taken to track NEAR Shoemaker's progress towards its goal. The asteroid got brighter, eventually bright enough to take a movie of its brightness variations as Eros rotated through one of its 5 hour 27 minute "days".

As we gradually came up on Eros, our plan was to take four giant mosaics of the space around the asteroid to look for moonlets, nearly a dozen movies of the asteroid rotating, and opnav images four times a day. Once in orbit, at any given time we would almost always be taking either some kind of mosaic or movie.

During all the excitement, Bob Farquhar and Dave Dunham ran a series of necessary but tedious meetings called "contingency planning." Bob said that, if anything went wrong, he wanted a plan in his back pocket, so Dave documented every conceivable mishap. NEAR needed four firings of its engine, each about a week apart, to match its speed with Eros and finally to go into orbit. The first was December 20. What if burn 1 was too small? What if burn 2 was too big? For each possibility, Dave had a stack of unreadable viewgraphs in microtype that showed exactly what needed to be done. Bob was particularly interested in what we would do if no burns happened at all. The celestial navigation team at JPL (Bill Owen and his colleagues Bobby Williams, Pete Antresian, Cliff Helfrich, Steve Chesley, and others) produced what was called a no-burn trajectory, and Bob told Ann and me to plan something in case that happened. Ann was too busy, so it fell to Maureen Bell (who had started working with Ann earlier in the year) at Cornell to crank out the SEQGEN. This would be like the Mathilde flyby: multiple views while flying by, except that the spacecraft would be going ten times slower, allowing lots more images, inbound and outbound satellite searches, a big mosaic at closest approach, and several dozen color images in between.

The big moment, the first and by far the largest burn of NEAR Shoemaker's main engine, was supposed to be at 6.00 p.m. on Sunday December 20. Mark Holdridge promised to send an e-mail to everyone on the project as soon as the burn happened. Science team members, various engineers, and others sat by their e-mail waiting for the word that the burn was a success. I sat in my office, five minutes from Mission Operations, on the off chance that anything came up involving the camera. The people who really mattered to the success of this maneuver were all gathered at Mission Operations. At 7:30 p.m. Ann Harch called and said she had talked to Karl Whittenburg, who had said that that burn had stopped after less than two seconds and that Mission Operations had lost contact with the spacecraft.

Somberly, Ann and I dragged out the contingency plan and started going over it. If nothing more happened on NEAR, the spacecraft would coast past Eros on the 23rd, alive or not. We planned on it being alive, taking on faith that Mark, Karl, Bob, and everyone else in Operations would figure out, somehow, how to save the mission. Right away Ann realized that the contingency plan Maureen and I had made up was no good any more; the engine's two seconds of burning probably threw us off course. We guessed by how much using calculators. We would need something new. We talked into the night until we were both too tired, and finally agreed that the next morning we would get down to work, adapting the basic observation plan Maureen and I had come up with to the new reality, whatever that might be.

On the morning of Monday December 21 there was still no word from the spacecraft. Everyone on the project wanted to help out in some way, but the only people who made any difference at this point were all in Mission Operations, sleepless. The unspoken rule was to stay out of their way – unless someone called for anything, and then to drop everything and do whatever was requested. Ann, Maureen, and I plugged away at the new contingency plan.

On the night of December 21 Ann called to say that she had spoken to Karl. Operations was back in contact with the spacecraft! After a 27-hour nightmare, the autonomy rules in the spacecraft's computer had managed to regain control so that NEAR Shoemaker could phone home. Again, we all talked late into the night and agreed on what to do in the morning.

On Tuesday, December 22, Bob Farquhar said to go ahead and finish the contingency plan. Operations was working on how to fire the engine again to match speed with Eros, but not before the 23rd. So take the pictures. All day the three of us worked frantically, Ann in charge, Maureen and I doing whatever she asked. We had a science team teleconference to apprise everyone of what was happening. At 4.00 p.m. there was a closed-door meeting of key spacecraft and instrument people, where they were told what Operations knew. NEAR Shoemaker was alive and working. It had spun out of control with its solar panels off the Sun for a long time. The battery (for emergencies only) had nearly drained but autonomy ruled! The low power had caused the recorder to shut down so the spacecraft could not remember what it had been through. The fuel tank pressure gauges showed that 28 kg of fuel were gone, and the few remaining records of what had happened suggested that NEAR Shoemaker's small attitude control thrusters had fired 600 000 times. We were 3000 km off course. The JPL navigation team provided an estimate of where we were and where we were heading. We would pass over the night side of Eros 4000 km away, but we could get pictures on the way in and the way out – distant ones.

By 9.00 p.m. Ann had finished the MSI sequence. She, Maureen, and I got on the phone and reviewed every letter of every command. It was ready for Karl. MSI, NIS, and the magnetometer would all take measurements as we coasted by Eros. The three instrument scientists, including myself, sat in Operations until 2.00 a.m. We were lucky to get to go home and to bed. Karl worked through the night to finish all the inputs, which were sent to the spacecraft with eight minutes to spare before the sequences were set to start at 10.00 a.m., December 23. The equivalent of the Mathilde flyby, which took six months of intensive work to sequence by hand, had been done and uploaded in two days under extraordinary circumstances.

At 6.00 a.m. on December 24, the scientists involved in MSI, NIS, and the magnetometer began to rendezvous at APL. Others remained at home, staying in constant touch over the internet. By 8.00 a.m. images were coming down. Pretty soon more and more views hit the ground. As each one appeared, increasingly excited voices blurted out "Wow!" "Would you look at that?" "Boy is that weird looking!" "Who took that big bite?" Eros's shape could be described as a bent peanut, a Dutch wooden shoe, or an Idaho potato with a giant bite taken out of one side. In the early afternoon, during a teleconference with the science team and project management, it was decided to get the best views on the Web by 3.00 p.m. so that, by Christmas Eve, the world could see that NEAR Shoemaker was one tough little spacecraft and could deliver a sleighful of images even after a near-death experience.

After that, people took breaks for Christmas dinner, but the workplace looked nothing like it should have done on a holiday. Most of Christmas Day, Clark Chapman and Bill Merline dutifully examined the pictures for Eros moonlets. They would find none. At Cornell, Jonathan Joseph (who works with Peter

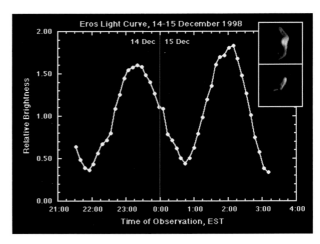

Figure 3.13. Variations in Eros's brightness were measured on December 14–15, 1998, when NEAR Shoemaker was at a distance of 460 000 miles (740 000 km). The period of the measurements covers just over one rotation of the asteroid. At this time, Eros still appeared smaller than one pixel in size, so its shape was not resolvable. The two inset images, acquired as NEAR Shoemaker passed the asteroid on December 23, show the approximate orientation of the asteroid at the points of maximum brightness (top) and minimum brightness (bottom).

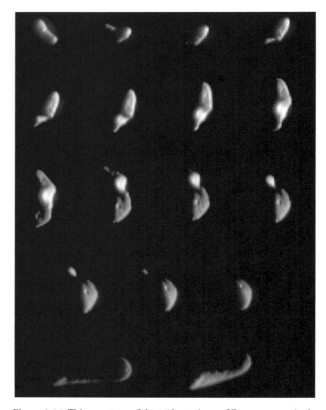

Figure 3.14. This montage of the 17 best views of Eros was acquired between 10:44 a.m. and 2:05 p.m. EST December 23 as the spacecraft range closed from 7300 miles (11 100 km) to 2320 miles (3830 km). From upper left to lower right, progressively less of the day side of Eros is visible as NEAR Shoemaker's view shifted from the day side to the night side.

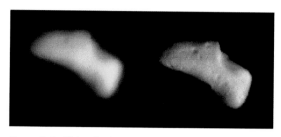

Figure 3.15. This color image of Eros, taken on February 12, 2000 as NEAR Shoemaker was approaching, is shown before (left) and after (right) application of the deblurring filter developed from images of the star Canopus.

Thomas) would make the images into the first flyby movie. And most importantly of all, Dave Dunham set to work on a new engine burn while Mission Operations worked to understand what had happened to NEAR Shoemaker, in preparation for a late maneuver and backtracking to Eros.

The year after

On the morning of December 26, Clark Chapman set the tone for what we would be doing over the next year. "There's a lot more scattered light in these images than there used to be," he announced. Sure enough, comparing images taken before and after "the anomaly" (the term that came to describe the events of December 20–21), the images had degraded remarkably. A faint halo surrounded Eros and the relative sharpness of the Mathilde images was gone. It was like looking at Eros through slightly fogged glasses. The next day, all team members were at APL for a meeting and we discussed it. The most optimistic idea was that water vapor (a product of those 600000 thruster firings) had condensed as a thin frost on the camera's outer lens and that, surely, by now it was gone. Perhaps we were hiding from the awful truth that back in 1996 we had blown the lens cover too soon and that we should have left it on until now because it was meant to keep contaminants like these off the camera.

A collective sigh of relief was breathed across the mission on January 3, 1999, less than two weeks after the anomaly. NEAR Shoemaker's engine was fired by just the right amount to match speed with Eros, at a half million kilometers away from the asteroid. Bob Farquhar announced that enough fuel was left to do the orbital mission, provided we conserved it by crawling back to Eros slowly, at highway speed. It would take a year, but we would get there – on Valentine's Day, 2000.

At another team meeting on January 5, we all took stock of what we had learned. Eros was 33 km long and 13 km wide, a little smaller than expected. Any variations in the asteroid's color or spectra from place to place were weak at best. However it clearly did have bright spots: the floor of the big bite in Eros's shape (by now termed "the saddle") was a good bit more reflective than the rest of the surface. Unfortunately, the flyby had been too distant (over 3800 km away) to see any small-scale details. A mere five craters could be counted. A long ridge running down the asteroid's spine was at the edge of resolution. The distance of the flyby conspired with the blurriness of the pictures to hide all of Eros's story except the generalities. Don Yeomans did produce a mass estimate from the radio tracking, and this was combined with the MSI volume estimate to yield a density estimate consistent with typical slightly porous common meteorite types, albeit with large uncertainties. We all agreed that we had learned a little about Eros but that the main value of this new knowledge would be that we now knew enough to customize the orbital mission to what Eros was really like. We would not be orbiting a body no-one had ever seen before.

Soon, though, attention came back to the health of the camera. Ever since the flyby, Bill Owen had been tracking the position of Eros using optical navigation images, and he was concerned that the asteroid was a lot dimmer than it had been during the approach. At first, it seemed that the weirdness of the shape might be responsible. We were over the night side, and at Mathilde that asteroid's strange shape placed an unduly large fraction of the surface in shadow from that perspective. But on February 5, the reason for Eros's persistent dimness became clear. One of the sets of optical navigation images by chance caught a bright star, and it was surrounded by a fat halo like a comet. The contaminants on the camera's optics were still there.

We needed a plan, right away. The "new" orbital mission was a year away, and the camera was dirty. We took two tracks at the same time. The first order of business was to figure out if the contaminants were stable, or were slowly evaporating. So we started an observation campaign of Canopus, the second brightest star in the sky and the easiest bright one to image from NEAR Shoemaker. Once per month through July, 16 images were taken through each of the camera's filters. Each time, we would measure the halo and look for any change. There was none. The contaminants just sat there. In parallel, Ed Hawkins and Rob Gold looked into what the chemical might be and whether there was a way to get rid of it. It turned out that the most likely materials were all very non-volatile and, even if they could be made to evaporate a little, there was nothing we could do that did not risk making matters even worse. After lots of hand wringing, Rob, Ed, Joe Veverka, Bob Farquhar, and I all decided that the best course of action was to do nothing drastic that could get us into worse trouble, and to focus instead on image-processing techniques.

Scattered light as a problem was not unknown on planetary missions. The imager on the Galileo spacecraft had a scattered light problem and, in that case, some of the scientists had used an image-filtering technique to try to remove the scattered light mathematically. Basically, the Canopus images were used as examples of how light from a single pixel is scattered around an image, and the filtering technique "packed" scattered light back into the pixels where it belonged. By October, the "final" method was in place and had been demonstrated on flyby images. Joe Veverka was so pleased that he asked for all further images of Eros to be deblurred using this technique.

While some of us were working on the deblurring filter, Ann Harch and Maureen Bell were busy replanning the new orbital mission. They were joined at Cornell by Colin Peterson, who had helped as an undergraduate assistant and proved so adept that he was hired after graduation for a year as a full-time sequencer. After NEAR Shoemaker's brush with extinction, the planning philosophy was to make the imaging plan as bullet-proof as possible. Nearly every observation was planned to be made no less than twice, at least a week apart in case the spacecraft was in safe mode or in the wrong position during the first run. Many imaging sequences were planned to take place four or more times. This brute-force approach was possible because the "new" orbital mission would take place while Eros was much closer to Earth than in the original plan, affording much higher data rates during DSN transmission. The original, pre-launch plan for NEAR was 40 images per day; when the orbital mission actually happened, the average number of images per day was nearly 500.

And everything was practiced, again and again. In contrast to the first approach to Eros a year earlier, the plan for the second go was incredibly well thought out and the sequencers were "old hands" at SEQGEN. Representative weeks taken from throughout the year were practiced on ground, and on December 15 and 20 the key observations just prior to orbit were practiced inflight. The perfect performance during the second practice was in happy contrast to the horrors of "the anomaly" exactly one year earlier. As the old millennium closed, everyone on the project looked forward to beginning the new millennium with the first orbital investigation of an asteroid.

An asteroid for the new millennium

On January 10, 2000, the final approach to Eros started, this time for real. The search for moonlets was repeated, for thoroughness, despite the lack of results at the flyby. The first three weeks were a slow-motion repeat of the approach in late 1998, except that for the fact that the seasons on Eros had changed. Back then we saw the southern hemisphere illuminated; now we saw the northern hemisphere. At first the asteroid remained a tiny blurry peanut. By February 1 it was clear that opposite the saddle lay a giant, round crater more than 5 km across. On February 6 the apparent size of the asteroid exceeded that at the flyby, and every couple of days a new level of detail became visible. Beginning then, each day a particularly cool new picture was posted on the Web with a caption – NEAR's so-called "Image of the Day." We started this with the intention of bringing the public along with us, and it worked. Over the full course of the mission the Image of the Day would draw 50 million hits.

The climax to the approach came on February 12, NEAR's megamovie, which we called the Multispectral Rotation Sequence. One frame, every half-degree of Eros's rotation. Every 12° of rotation there was a full color sequence to map the surface composition. Unlike special-effects asteroids in Star Wars, this was an actual movie of a rotating asteroid. NEAR had arrived.

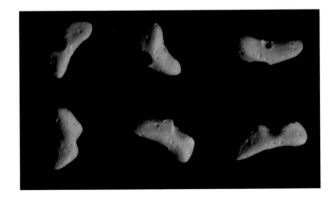

Figure 3.16. These color images, taken February 12, 2000 during NEAR Shoemaker's approach, show different parts of the northern hemisphere of Eros as seen from the spacecraft over one full rotation of the asteroid.

4 Landscape of an Asteroid

Louise Prockter
Applied Physics Laboratory, Johns Hopkins University (NEAR Science Team Associate)

Mark Robinson
Northwestern University (NEAR Science Team member)

The NEAR mission to the asteroid 433 Eros has offered planetary geologists an unprecedented opportunity to study the surface of an asteroid in great detail. The data collected are already leading to huge advances in our understanding of the processes that shape and modify Eros and other asteroids.

Eros is not the first asteroid that a spacecraft has visited. The Galileo spacecraft flew past main-belt asteroids 243 Ida and 951 Gaspra in the early 1990s en route to the Jupiter system, and NEAR Shoemaker itself swooped by the large dark main-belt asteroid 253 Mathilde in 1997 (see Chapter 3). However, NEAR Shoemaker was the first spacecraft to orbit an asteroid, returning images of the entire surface with better than 5-m per pixel resolution, and eventually sending back its final image with a resolution of only 1.2 cm/pixel just before setting down on the surface. During the year NEAR Shoemaker was in orbit, its camera, the multispectral imager (MSI), snapped over 160 000 images of the surface, many in color. The spacecraft obtained data over a wide range of lighting conditions, zoomed in and out, and studied every nook and cranny on the asteroid. Eros is now one of the best-imaged bodies in the solar system – perhaps even more so than our Earth, two-thirds of which is covered by ocean and therefore defies close scrutiny.

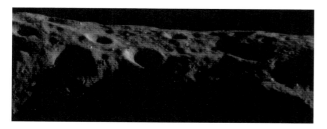

Figure 4.1. An oblique view along the terminator of Eros from optical navigation images acquired on August 6, 2000 while NEAR Shoemaker was in a 50-km orbit above the asteroid.

Eros has emerged as a surprisingly complex body. Like most other objects in the solar system, impact cratering has clearly shaped the asteroid at all scales. Its surface is crisscrossed with a network of fine fractures and troughs, and a very long, prominent ridge winds around much of its mid section. Rocks the size of anything from a chair to a house are strewn everywhere. Inside large craters we see evidence of landslides that have exposed fresher material. Close up, we find smooth ponded deposits of fine-grained material filling some craters and depressions. This chapter will discuss the overall landscape of Eros and the details of the different geologic features observed by NEAR Shoemaker. Then we will present some possible explanations of how Eros has come to be this way.

Cartography

Mapping a regular planetary body is challenging enough, but Eros introduces a whole new level of difficulty. How do you make a map of something the shape of a potato? Where is north? Where is the equator? How do you put that map down on paper so other researchers can easily locate features on the surface?

The first step in this process is the creation of a computer model of the shape of the asteroid. This was done using knowledge of the position of the spacecraft (measured using the Doppler shift of its radio signal as it orbited Eros), the position of stars seen by a special on-board camera, and with MSI images of landmarks on the surface. By tracking the locations of features from different orbits and different viewing angles, a three-dimensional model of the asteroid could be built up as new images were received from the spacecraft. As more and more images were obtained, and more of Eros was observed, this "shape model" became more detailed. The final shape model is a very close approximation of Eros, and only differs at resolutions smaller than about 1° in latitude and longitude (see Chapter 5).

Figure 4.2. A global map of Eros constructed from more than 300 separate images obtained by the MSI instrument. All longitudes spanning latitudes from 60°N to 60°S are shown in a simple cylindrical projection (bottom). The north polar region (upper left) and south polar region (upper right) are shown in polar stereographic projections.

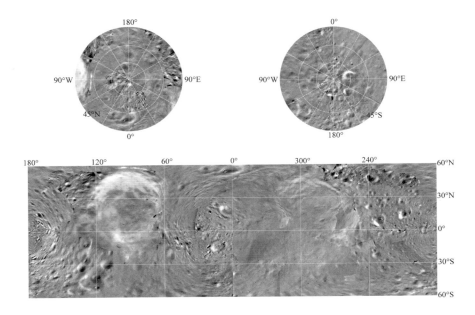

Figure 4.3. Six equatorial and two polar perspective views of Eros. These views were produced by overlaying the global base map on top of a digital model of the shape of the asteroid.

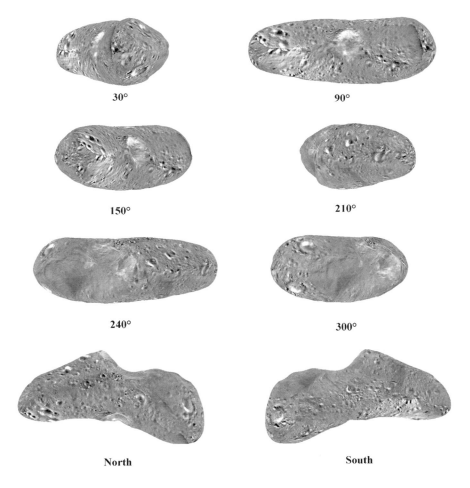

Figure 4.4. A global base map of Eros from 90°N (top) to 90°S (bottom) with proposed names of major craters superimposed. Following the conventions of the International Astronomical Union, features on Eros are named after prominent lovers from history, literature, and mythology around the world.

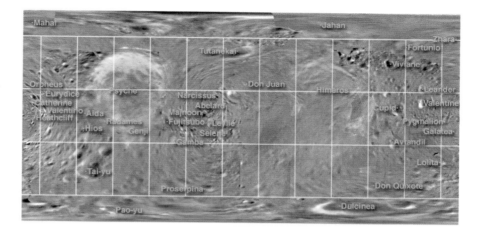

Once the shape model was generated, images obtained by NEAR could be overlaid on to it to create a three-dimensional model of Eros.

The position of the equator on Eros is designated the same way it is on Earth; that is, it is the plane that is perpendicular to the asteroid's direction of spin. On Earth, the line of zero longitude, or prime meridian, was chosen in 1884 to pass through the site of the Royal Greenwich Observatory in England. On Eros, the prime meridian was defined to pass through a prominent crater that was visible during the 1999 flyby on one of the long ends of the asteroid. Once these two points of reference had been established (0° latitude and 0° longitude), the rest of the latitude/longitude net was determined from the shape model.

Presenting data in map form is far easier for a spherical body than for something with an unusual or irregular shape, but even then the map projection must be chosen carefully to minimize distortion. Eros is shown in a simple cylindrical projection in Figure 4.2. Conceptually, this projection can be thought of as being generated by "unzipping" the surface of Eros along the prime meridian and rolling it out on to a rectangular sheet of paper. This is a common projection used for many global maps of the Earth. A disadvantage of this projection is the extreme stretching out of regions near the poles. For a similar map of Earth, the scale is constant from longitude to longitude as you travel around the equator, but for Eros the scale changes by a factor of two around the equator because the asteroid is roughly twice as long as it is wide. This means that these maps of Eros have twice the distortion problem as maps of spherical planets such as Earth. The best way to handle this problem is to take the map and reproject it on to the shape model to produce a natural view as seen from an observer at any desired point above the asteroid. These

are often called point-perspective views. This is a powerful and intuitive way of visualizing the surface. For example, by making hundreds of point-perspective views, each with only slightly varying viewpoints, a movie simulating a flight over the surface can be generated.

Naming of features

When a planetary body is imaged for the first time, scientists collecting the data and members of the International Astronomical Union (IAU) choose a theme for naming features on the surface. Typically, names may be taken from mythology, literature, or something else relevant to that particular body. For example, features on Venus are named after goddesses and famous women; those on Mars's moon Deimos are named after authors who wrote about the martian satellites; and those on Mathilde are named after coal fields of the world (due to the asteroid's dark nature and presumably high carbon content). Generally, scientists propose specific names for features they are studying within the theme determined by the IAU. Once accepted, the names can then be used on official maps and other publications. In addition to the naming theme, each type of feature on a surface has a name or class of names which is common to landforms of that type on any planetary body. For example, a ridge is referred to by the Latin name "Dorsum" (for example, Rahe Dorsum on Eros), while "Mons" signifies a mountain (for example, Olympus Mons on Mars).

Since Eros was the god of love in Greek mythology, an appropriate theme for the asteroid of the same name is famous romantic figures. One of the largest craters is named Psyche, with whom Eros was in love; another is named Narcissus, who fell in love with his own reflection. Other craters are named after romantic figures in literature, such as the 17th-century French paramour Don Juan, and Genji and Fujitsubo, lovers in "The Tale of Genji," written around 1000 AD and arguably the first modern novel. Additionally, features are named after key figures in Eros's discovery and exploration. One example is Shoemaker Regio, a large, boulder-strewn region named after the late Eugene Shoemaker, one of the founders of planetary geology and co-discoverer of over 800 asteroids. A second is Rahe Dorsum, named after the late Jürgen Rahe, a planetary scientist and comet expert who played a key role in making the NEAR mission a reality.

Illumination effects

Eros is slowly rotating on its spin axis once every five hours and 16 minutes, and it completes one orbit around the Sun every 1.76 Earth years. The spin axis of the asteroid is tilted almost horizontally – much more than Earth's – so the asteroid is slowly spinning end over end in the same way a thrown shoe would travel through the air. Because of this extreme tilt, the seasons on Eros are considerably exaggerated. Earth's spin axis is tilted by 23.5° from the direction at right angles to its orbit, and so the northern hemisphere experiences longer days in summer, and shorter days in winter. Eros's spin axis, with a tilt of 88°, is virtually in the same plane as its orbit, resulting in almost total daylight in summer and complete darkness in winter for a whole hemisphere; you could say that it is the ultimate land of the midnight sun. NEAR Shoemaker

first reached Eros in the northern hemisphere's summer season; the southern hemisphere of the asteroid was in darkness until it slowly became illuminated during the course of the mission. It was very exciting when new and unexpected features were revealed as the southern hemisphere gradually lit up. By the end of the mission, the northern hemisphere was almost completely in darkness and the southern hemisphere was fully sunlit.

One of the most challenging problems facing geologists when looking at planetary surfaces is the effect of variations in lighting on a scene. Shadows change in length depending on what time of day it is, so it is imperative for geologists to understand the conditions under which an image was taken before trying to estimate the height or depth of a landform based on shadows. Figure 4.6 shows a region near the south pole of Eros, imaged at four different times up to seven weeks apart. On June 27 the area looked fairly smooth, with a few apparently shallow craters and some very faint lineations running across the scene. With the change in illumination, the lineations trending NE–SW across the scene became much more apparent. When the Sun was low on the horizon, features like troughs were very prominent and the craters much more distinct. These images clearly show that, when making a map of any planetary body, each area should ideally be looked at under as many different kinds of lighting as possible. Often, though, this is not feasible. For example, when NEAR Shoemaker flew past the asteroid Mathilde, only a small number of images could be taken in the time available, and then only of

Figure 4.5. Because the spin axis of Eros is tipped nearly on its side, the asteroid experiences extreme seasonal changes in solar illumination during each 1.76-year trip around the Sun.

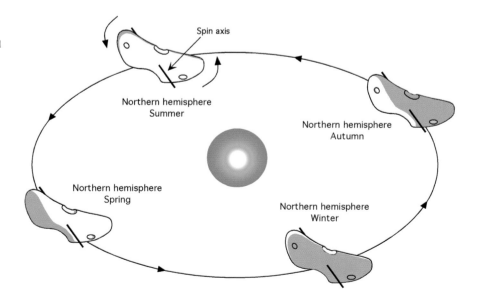

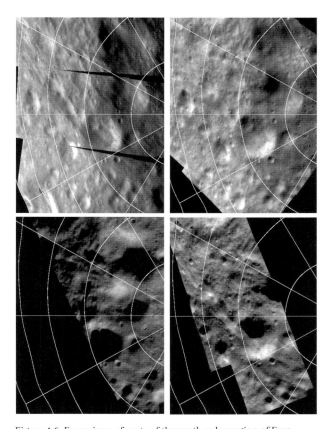

Figure 4.6. Four views of parts of the south polar region of Eros, seen under different lighting conditions, corresponding to different times of day. The Sun is highest in the sky in the top left panel and gets lower in the sky in successive panels, clockwise around the montage. Notice the longer shadows in the lower-left panel. The topography appears much more subdued when the Sun is higher in the sky (upper panels) than when it is closer to the horizon (lower panels), emphasizing why NEAR's feature mapping objectives were achieved primarily from "low-Sun" imaging. Curved gridlines are lines of latitude; straight lines are lines of longitude, which converge at the south pole just off to the right of each image.

one side of the asteroid and with very little variety in lighting conditions. We have been extremely fortunate that NEAR Shoemaker was able to orbit Eros for an entire year, imaging the same features at many different angles. This ability to see the individual features on the asteroid over and over again under different conditions has greatly increased our understanding of their morphology (shape and surface roughness) and formation conditions.

Main morphological/structural features

Craters
Craters are the most common landform in our solar system, especially on bodies with no atmospheres.

Because craters are formed at predictable rates that were higher in the past, we can infer that the more craters an object has on its surface, the older it is. The impact crater density of a surface (number of craters per square km) can also be changed if the body is resurfaced in some way, for example by volcanic flows or wind or water erosion covering up some craters. Based on the characteristics of its orbit, many astronomers believe that Eros actually spent much of its life not as a near-Earth asteroid but as a member of the main asteroid belt between Mars and Jupiter. The main belt is a region that at one time early in the history of the solar system would have had many more chunks of debris colliding with one another than the region of the solar system near the Earth. The similarity between the density of craters on Eros and the crater densities of other main belt asteroids (Gaspra, Ida, Mathilde) bears witness to this violent history. Eros is not big enough to have had a molten interior, and therefore has no volcanoes on its surface, nor does it have an atmosphere or water for weathering and erosion to operate. The most important process acting to shape the surface of Eros is impact cratering, and we can assume that most of the craters – at least the largest ones – are preserved on the surface, unless they have been modified by later impact events. Since objects tend to break up when they collide, the biggest craters were probably formed early in the history of the solar system, when pieces of debris were larger and had not yet been broken down themselves by collisions. Small craters are probably still being formed today on Eros, although at an extremely slow rate.

The craters on the surface of Eros have a variety of shapes and morphologies. Some appear very old and degraded so that they are little more than barely discernable depressions in the ground. Others appear relatively fresh (bear in mind that in geological terms something that is "relatively fresh" could have formed only a billion years ago, instead of 4 billion), with crisp rims and sharp outlines. Larger craters on Eros commonly have bright material on their steepest walls (above 25°). When we saw the first images taken by NEAR Shoemaker, we thought that the walls were bright simply because they were reflecting the sunlight more directly. However, when we viewed the craters from different angles and under varying lighting, it soon became apparent that the walls really were intrinsically brighter than the surrounding terrain (having a higher albedo, in astronomical terms), suggesting they either had a different composition, were made of different sized grains, or were

Figure 4.7. A view of typical cratered terrain on Eros obtained from optical navigation images acquired on April 24, 2000 while NEAR Shoemaker was in a 100-km orbit above the asteroid. The image measures approximately 5 km from top to bottom.

younger. From lunar exploration we know that as rocks on an airless body are exposed to space they darken and their color changes. The most likely processes responsible for this darkening are the constant bombardment by cosmic radiation and the solar wind, and high velocity impacts by tiny rock and dust particles called micrometeorites (see Chapter 6).

The 5.3-km crater Psyche is large enough that even from high orbits we could take a good look inside it and examine the striking bright patches and streaks on its walls (see Figure 6.8). We discovered that loose material seems to have sloughed off the walls and moved downwards, collecting on the crater floor in a manner similar to that of terrestrial landslides. The removal of the loose material reveals younger, apparently less-weathered, brighter material on the crater walls.

In addition to Psyche, Eros has two other large impact craters on its surface. Himeros was the most conspicuous landform apparent in the 1999 flyby images of Eros. This feature does not look like a classic impact crater, which generally has a rounded shape

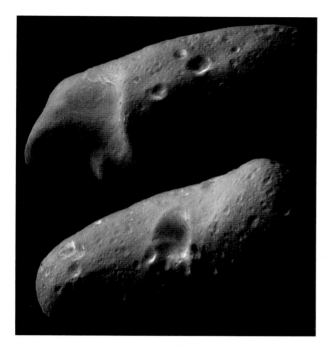

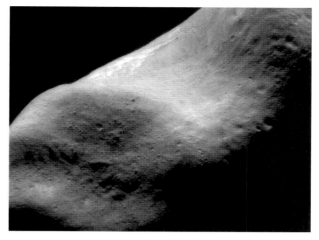

Figure 4.9. Mosaic of the large, smooth crater/depression Himeros (background, top) and the smaller rough-textured crater Shoemaker (foreground, lower left). Images were obtained on April 10, 2000 from 100 km orbital altitude.

Figure 4.8. Mosaics of the eastern half (top) and western half (bottom) of the northern hemisphere of Eros, obtained on February 23, 2000 from a range of approximately 355 km (220 miles). The top view is dominated by the large crater/depression Himeros (left side), and the bottom view is dominated by the 5.5-km-diameter impact crater Psyche (center). The smallest details visible in these mosaics are 35 m (120 feet) across.

and a raised rim, but instead is shaped like a saddle. Part of the reason for this is that Himeros is very large (9.4 km in diameter) relative to Eros itself, which measures 13 km across its midsection. Thus, any big impactor (an asteroid or comet) hitting Eros at this narrow point will make a different shaped crater than if it were hitting a relatively flat surface as it would on most larger planetary bodies. Although the actual process of making the crater is different, you can get an idea of how this works by imagining the imprint you would get by throwing a tennis ball at a tube of putty the size of a soda can. When viewed from certain angles and lighting conditions, a rim is clearly evident on Himeros's eastern side, betraying its impact origin. Its southwestern rim has been obliterated by the formation of a younger crater, called Shoemaker, which is 7.3 km in diameter.

Himeros has a lightly cratered interior filled with grooves and troughs of a variety of sizes and shapes, but much of it is very smooth. Since Himeros appears to be old, we would expect its interior to have many craters on it, but this is not the case. Where have most of the Himeros craters gone? One possibility is that they have been filled and obliterated by landslides tumbling down from the walls.

Figure 4.10. This view of Eros shows the inside of the 10-km-wide depression/crater Himeros, and was acquired on August 5, 2000 while NEAR Shoemaker was in a 50-km orbit above the asteroid. The lighting reveals a number of linear markings that may represent tracks left when boulders ejected by impact events bounced across the surface. Some of the circular and semi-circular features on the walls of Himeros appear to be impact craters which have been partially buried by ejecta from later impacts, or by loose material moving down Himeros's slopes. The largest of these features measure about 1 km across.

Boulders

The crater Shoemaker is unique among Eros craters in that it has an unusually high density of boulders and loose rock fragments lying within it. This heavily boulder-strewn area is called Shoemaker Regio. Scientists have developed models to predict the

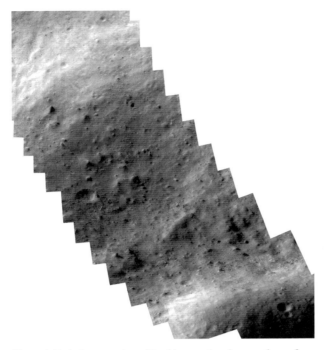

Figure 4.12. Examples of partially buried boulders and enigmatic bright mantling deposits (lower left of center) in a high-resolution image of Eros. This image is about 300 m across.

Figure 4.11. A close-up view of boulders scattered across the surface of Shoemaker Regio. This mosaic was acquired from a 50-km orbital altitude in May, 2000. The brighter areas at top and bottom of the image are the rims of Shoemaker crater, which are about 7 km apart.

expected paths of dust and boulders (known as ejecta) thrown out during an impact, to find out where they might land. Modeling of ejecta trajectories from the formation of Shoemaker crater suggests that most of the boulders lying on Eros's surface today may have been thrown out during the excavation of that one crater. Boulders within Psyche crater are probably from the Shoemaker crater impact event, implying that Shoemaker is younger than both Psyche and Himeros, and is probably the most recent large crater on the asteroid.

Boulders on Eros have a multitude of sizes, ranging from less than a meter across to the size of whole city blocks (150 m or 500 feet). They can be angular or rounded, and close up some look weak and friable (crumbly). Many boulders are clustered together inside craters or other depressions. They have probably been shepherded by Eros's weak gravity so that they are resting in topographic lows, just as most boulders rest on the floors of valleys on the Earth. While many boulders are completely exposed above the ground, some appear to be embedded, or partially buried, within it. Some typical examples are in Shoemaker Regio; some appear almost pyramid shaped, with their lower parts completely obscured. These blocks may be partially covered by deposits of dust and rock from

later impacts. Figure 4.12 shows a large rock standing on a smooth patch of material. Surrounding the rock is a narrow mantle of bright material that appears to have been shed from the rock on to the adjacent terrain.

Ponded deposits

When NEAR Shoemaker swooped within 5 km of the surface of Eros in October of 2000, we obtained many spectacular high-resolution images. Some of these images showed unusual smooth areas that we had not been able to identify from higher orbits. In addition to having a smooth and unbroken texture, these areas are very flat, and are only found inside impact craters and other depressions. Because of their appearance, they were informally termed "ponds" by the science team, even though water or other fluids have probably played no part in their formation. The flatness of the ponded deposits implies that they may be composed of relatively small grains of material, centimeters or less in size. Since ponds are found in topographic lows they probably formed when relatively fine-grained material moved down the slopes of the surrounding terrain and settled on the floors. The ponds are probably only a few meters deep, based on morphologic clues and the sizes and depths of their host craters. Nothing like them has been seen on any of the other asteroids or small bodies studied to date. It is interesting that most of the ponds seem to be clustered around Eros's

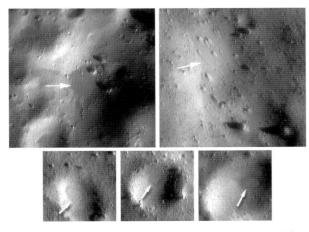

Figure 4.13. Examples of "ponded" deposits of flat-lying materials (arrows) in high-resolution images of Eros. The image at the top left is of an area 800 m across; the image at the top right is of an area 200 m across; both bottom images are of areas 120 m across.

equator. At this time, we do not know for sure why some craters and depressions contain ponded deposits and others do not, but we can speculate.

The ponds may result from a process called "seismic shaking," which is important for small bodies like Eros and which may also explain why some of the craters in Himeros appear to be "missing". Seismic shaking occurs on Eros whenever there is an impact somewhere on the asteroid. The shock, or seismic, waves from the impact may travel across the surface and through the body of the asteroid, the distance traveled depending on the size of the impact and how fractured or broken up the interior is. Similar behavior is seen with seismic waves caused by terrestrial earthquakes. For example, the great San Francisco earthquake of 1906 was felt as far inland as central Nevada, and from southern Oregon to south of Los Angeles along the coast.

One effect of seismic waves can be to cause loose, fine-grained material on the surface to move down hill and level out in depressions. A commonplace example of seismic shaking happens in kitchens all over the world on a daily basis. One of the first steps when baking a cake is to sift flour into a bowl. At first the flour piles up into a mound below the sieve. If the bowl is sharply struck on the side several times, the heap of flour gradually levels out. In an analogous process, the Eros ponds may have formed when relatively fine debris tumbled down slopes, then leveled off as seismic waves from later impacts had the same effect on the surrounding rocks as does tapping a bowl of flour. The larger the grains, the harder it is (and the

greater the shock needed) to level the debris, which is consistent with ponds being made of relatively fine material. However, there are some problems with the simple seismic shaking idea. Mainly, how did the fine material that forms the ponds get into the craters in the first place? Perhaps a process not generally found on the Earth is responsible. Apollo astronauts were able to observe that extremely fine dust on the Moon was moved about the surface by electrostatic effects. As the Sun set, or rose, the boundary between night and day became electrically charged a tiny bit, just enough to cause fine dust to move. A similar process may be at work on Eros, and scientists are now trying to unravel this vexing problem.

Ridges and the "Twist"

One of the most remarkable landforms on Eros is a giant ridge that stretches for 18 km around the narrow midsection of the asteroid. The ridge, Rahe Dorsum, is so large that it was visible when NEAR Shoemaker flew past Eros at a distance of about 3800 km in 1999. It stands a few tens of meters high but has relatively shallow slopes. The ridge has a different cross-sectional shape at various points along it. One end is inside Himeros, where it changes from a ridge into a series of wide, flat-floored troughs. The ridge trends northwest up across the northern hemisphere and down the other side, terminating east of Psyche. Given the size of Eros, Rahe Dorsum indicates that some global process must have acted at one time and deformed the asteroid.

Near the north pole, Rahe Dorsum has an asymmetrical shape, with a smooth, low slope on one side, but a very steep, rough face on the other. On Earth, this kind of landform shape can occur when the ground is fractured during an earthquake or some other tectonic event, and the ground on one side of the tear fault becomes pushed up and over to the other side. The steep face of the upthrust block is now unstable, and landslides may occur. The end result is a feature that looks remarkably like Rahe Dorsum. For this reason, we think that some cataclysmic event, probably a large impact, occurred on Eros, perhaps causing a fracture plane to form right through the asteroid, and resulted in one half being thrust up over the other by a few tens of meters.

We do not yet know which of the impact craters on Eros might have been responsible for causing Rahe Dorsum to be thrust up. It must have been a large event to cause so much mayhem. We know it was not Himeros, since the ridge starts within that impact

Figure 4.14. Top: Examples of the
ridge named Rahe Dorsum (arrowed)
taken from a 100-km orbital altitude.
The large depression at the bottom is
Himeros. Bottom: Cartoon sketch of
the possible origin of Rahe Dorsum as
a thrust block forced upwards by
compression and faulting of the subsurface.

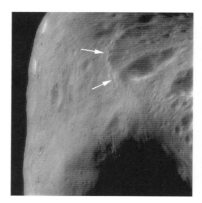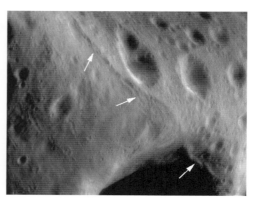

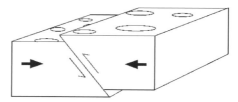

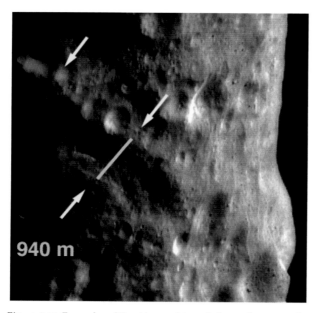

Figure 4.15. Examples of the ridge-and-trough feature known as the
twist in images taken from a 100-km orbital altitude. The ridges
shown here by arrows bound a trough nearly 1 km wide.

through which troughs – we are beginning to piece together the history of the asteroid and to understand how it got to look the way it does.

Another piece of evidence suggesting that Rahe Dorsum cuts right through the asteroid comes from a large ridge-and-trough set in the southern hemisphere. This feature, informally called "the twist," has provisionally been named Calisto Fossae. The troughs are several kilometers long and many hundreds of meters wide, so they are large features for a small body like Eros. Their location on the asteroid gives them the appearance of threads on a screw, leading to their informal name of "the twist." The ridges and troughs that form the twist region appear very ancient based on the many superimposed impact craters and the generally subdued and eroded topography; such an appearance implies a long history of battering by later impacts. The orientation of the twist is the same as that of the plane described through the asteroid by Rahe Dorsum, suggesting that the two features might be related. Since the twist appears to result from extension (pulling apart) of the asteroid, and Rahe Dorsum on its opposite side appears to be a compressional feature, together they provide intriguing evidence that Eros underwent some sort of global-scale bending during its lifetime, perhaps as a result of the Shoemaker crater impact event.

Grooves and troughs
In addition to being peppered by impact craters, the surface of Eros is crisscrossed with a network of grooves and troughs. These range in length from a few

crater, so it must be younger; if it had existed prior to Himeros's formation, it would have been blasted clean away when that impact occurred. By examining how Rahe Dorsum cuts across some grooves that intersect Psyche on the other side of Eros, we can tell that it is younger than that impact as well. This leaves one of the other large impacts on Eros, probably Shoemaker (the youngest large crater), as the culprit. By looking at these kinds of superposition relationships – how one crater overlies another, or which ridge cuts

Figure 4.16. Grooves seen in a high-resolution and low-Sun mosaic of Eros taken from a 50-km orbital altitude. The grooves each measure about 250 m across and are located in the northern hemisphere on one of Eros's ends, at 50°N, 185°W.

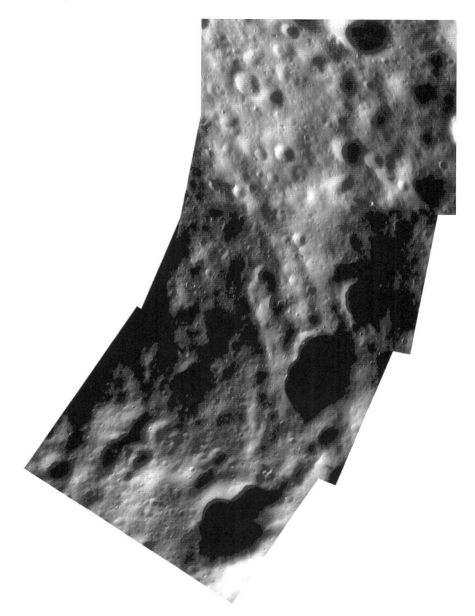

tens of meters to several kilometers, and may be up to a couple of hundred meters wide. Some have very straight parallel margins, while others have a scalloped or beaded appearance and look like long chains of small pits. Many of the troughs are almost filled in with debris (perhaps as a result of seismic shaking), while others look relatively fresh and may be tens of meters deep. Grooves can occur singly, or in sets of small troughs. They are often at, or near, right angles to each other. Some can be seen to cut across others, allowing us to establish which are younger.

Eros's grooves are of a similar size and appearance to those seen on the asteroids Gaspra and Ida, and also on the martian moon Phobos. The presence of grooves on the majority of the small bodies visited by space-

craft suggests that they are the result of a commonplace mechanism. The most likely suspect is impact cratering. When an impactor slams into a body, not only does it produce a crater, but the shock waves it creates may fracture the body further away from the impact site. These fractures may be covered up with debris from later impacts, or they may be exploited as zones of weakness (as may have happened for Rahe Dorsum).

Researchers studying grooves on Phobos, which also have scalloped margins, think that they represent surface traces of fractures lying beneath a layer of dust and debris called the regolith. In planetary terms, regolith is the name given to soils that are largely impact derived, unlike terrestrial soil, which is the end

Figure 4.17. An enigmatic 2-km-long curved trough within the crater/depression Himeros. This mosaic was obtained by the MSI instrument from a 50-km orbital altitude.

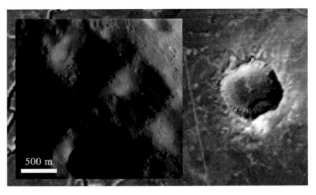

Figure 4.18. Square-shaped craters on Eros imaged from a 50-km orbit (inset) compared to the classic example of a structurally controlled craterform on Earth, Meteor Crater in Arizona, imaged by the Landsat satellite. The Eros craters are almost the same diameter as Meteor Crater, which is just over 1 km across.

product of weathering processes. Scientists who study Phobos think that the scalloping occurs when a fracture that has been buried undergoes a little further widening, perhaps as a shock wave from an impact passes across it. The regolith above the fracture becomes unstable, and drains down into the newly formed crevasse. For reasons that are not fully understood, the drainage process forms pits that are about the same diameter as the regolith is thick. These kinds of observations may allow scientists to estimate the depth of the regolith overlying the fractures. If this model is also applicable to Eros (and there is no reason why it should not be), the regolith overlying some fractures must be several tens of meters thick. Continued drainage of regolith and ongoing seismic shaking may act to fill in the pits, so the grooves become shallower over time.

One of the most striking troughs on Eros is a large curved, flat-floored trough inside Himeros. The trough begins where Rahe Dorsum ends, is about 2 km long, and measures over one third of a kilometer at its widest point. Many tectonic features – particularly landforms that occur as a result of large-scale movements of the ground, such as earthquakes – have straight, parallel margins, but the trough in Himeros

has a backwards "C" shape. Why is this? The answer comes from its location inside Himeros, the largest impact crater on Eros. The concave shape of the crater has almost certainly affected the trough, so that it has been forced to curve around inside the walls.

One interesting effect occurs when preexisting fractures and crevasses on Eros interfere with crater formation. Impact craters tend to be round in plan view (looking straight down) when they form on a surface that is essentially flat compared to the size of the crater. However, some of Eros's craters are almost square in appearance. This strange phenomenon occurs because fractures and crevasses that already exist in the asteroid have the effect of focusing the shock waves from a cratering event. Instead of moving out radially (in the same way ripples move out on a pond surface when you throw a stone into it), seismic waves are channeled along the fracture planes. This changes the resulting crater shape, so that it is influenced in the direction of the fracture orientations. Blocks are thrown out in a pattern matching the fractures, and the rim may have straight segments. A great example of this process is found on Earth at the Barringer (Meteor) Crater in the Arizona Desert, which has a somewhat square outline. The presence of slightly rectangular craters on Eros is further evidence that the asteroid is highly fractured close to the surface.

Descent imaging

On February 14, 2001, NEAR Shoemaker came to the end of its historic mission in spectacular style, by de-

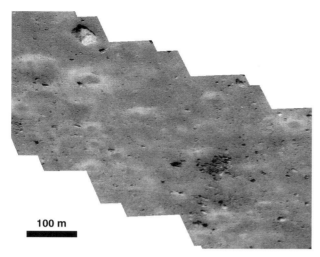

Figure 4.19. The first mosaic of MSI images generated as NEAR Shoemaker began its descent to the surface of Eros on February 12, 2001. The height of the spacecraft above the surface was about 5.5 km at the time these images were taken.

images down to an altitude of about 500 meters. Below that, scientists expected the images to be blurred due to the optical design of the camera, coupled with the speed at which the spacecraft was moving over the surface. Once again, however, NEAR Shoemaker surprised us, by returning incredible images down to only about 100 m above the surface. Amazingly, the very last image showed both a boulder and a pond, exceeding even our wildest hopes. It was during transmission of the final image that the spacecraft came into contact with Eros, causing the loss of the bottom portion of the last image as the antenna was jostled off its link with the Earth.

The descent images showed boulders in more detail than we had ever seen them before. We could see that some of them have come to rest on the surface in clumps, as if weak rocks had broken up upon impact. The surface has boulders down to the smallest scales. Some of the boulders are heavily fractured and look very similar to lunar boulders visited by the Apollo astronauts. On the right side of the final mosaic we see a rougher surface, and the rocks appear to be covered with a loose adhering layer of soil (small-sized component of the regolith). This area appears to be a steeply sloping wall of a crater – thus it appears NEAR landed in the middle of an impact crater about 100 m in diameter.

The last picture showed a distinct change in surface roughness about halfway from the upper right to the lower left. The lower portion of this image reveals the surface of a pond at incredibly high resolution (1.2 cm/pixel). This was extremely lucky; our hope had been to image, at some point in the descent, one of the ponds to see if they still look smooth at high resolution. Not only did we get a pond in the

orbiting and landing on the surface of Eros, and then continuing to operate for two more weeks. A combination of perfect lighting conditions and restrictions on the spacecraft's radio antenna pointing led the NEAR mission operations team to pick the interior rim of Himeros as the ideal site to touch the spacecraft down at the end of the mission. While descending to its final resting place on the surface, the camera snapped a series of images at the rate of two a minute. NEAR scientists particularly hoped to obtain high-resolution images of ponds, to see just how smooth they really were, and boulders, to see whether close up they looked like the boulders photographed on the Moon by the Apollo astronauts. They hoped to return clear

Figure 4.20. Mosaic of final descent images acquired by NEAR Shoemaker on February 12, 2001. The spacecraft appeared to make a transition from a relatively rocky region to a relatively smooth one, perhaps coming to rest on one of the mysterious, smooth "ponded" deposits identified on Eros. The images were obtained from right to left.

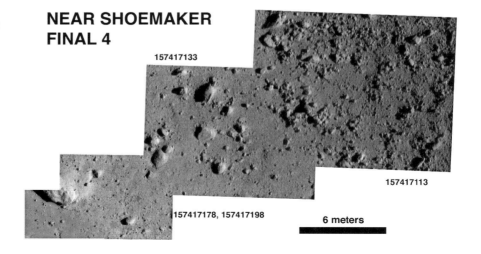

NEAR SHOEMAKER FINAL 4

157417133

157417113

157417178, 157417198

6 meters

last image, confirming that they are smooth at the 1-cm level, but calculations show that we probably landed right on that pond! Perhaps this is why NEAR Shoemaker survived the impact: by landing on a smooth surface within the crater we avoided drastically upsetting the spacecraft, and possibly the fine-grained materials that compose the ponds acted as a cushion (like landing on a sandy beach). But that is not the end of the story. In this last image we also see some very intriguing depressions right where the transmission was cut off. On other planetary bodies these kinds of depressions appear to be collapse pits, where the ground is porous and the surface has sunk down. As is commonly the case with images of other planets, one of the most interesting discoveries of the mission occurred during the last image, and only a few lines of the image containing it were received! Nevertheless, even this small amount of intriguing information is helping us to understand better more of the properties of Eros's regolith.

Implications for geological history

By studying the landforms on Eros we can better understand its violent past and perhaps more recent gentle geologic history. Impact cratering has clearly played a big part, creating both the numerous craters scattered about the surface, as well as forming the fractures along which troughs and grooves later developed. Cratering is also responsible for breaking down the surface rocks into smaller pieces of debris and dust, to form the regolith. Part of this regolith consists of numerous boulders strewn across the surface. In places, crater-wall material has sloughed off due to Eros's weak gravity, revealing fresher, brighter rock beneath, and resulting in landslides on the crater floors. Surprising evidence for very small collapse features in the highest-resolution images probably indicates that the regolith has a complex subsurface structure.

The largest craters on the asteroid – Himeros, Psyche, and Shoemaker – are all relatively old. Shoemaker is likely the youngest of the three, and its creation spewed a multitude of boulders across the surface. It may even have been responsible for the formation of Rahe Dorsum along a zone of preexisting weakness. And Eros continues to evolve geologically: surface processes such as mass wasting – the shedding of material down slopes – are probably still continuing today, every time there is an impact – however small – on this fascinating little asteroid.

5 Eros: Form and Substance

Peter Thomas

Cornell University (NEAR Science Team leader)

Although the NEAR Multispectral imager (MSI) and the Near Infrared Spectrometer (NIS) made direct measurements of only the very uppermost surface of Eros, one of the prime goals of the NEAR mission was to find some way to investigate its interior. There were a number of questions we wanted to answer. What is the overall composition of Eros? How porous is it? Are there large fractures running through it? Does the composition or porosity vary within the asteroid? Is the inside the same as the surface?

The inside from outside

Trying to infer what is inside from what is on the surface is not a hopeless task. Because of Eros's small size, the internal conditions of pressure and temperature are fairly uniform throughout. The inside does not heat up due to radioactive decay as does the Earth's. The pressure at the center of Eros is less than the atmospheric pressure around you as you read this book. Thus, the internal chemistry is not necessarily different from that of the surface, and cracks in subsurface rocks may not be closed by the pressure of the overlying materials, as they are on the planets and much larger asteroids.

The first indirect method of studying the interior of Eros was simply to determine and interpret its mean density (mass per unit volume). The densities encountered in solar system materials cover a wide range. Water and ice, for example, have a density of about 1 g/cm³. Typical rocks from Earth's continents and most of the common meteorite types have densities of around 3–4 g/cm³. Metals like iron and gold have densities upwards of 8–10 g/cm³. So just by determining this one number for Eros we could potentially find out whether the interior is rocky, or icy, or metallic. However, in practice the situation is somewhat more complicated, as it is also possible for a small body like an asteroid to have a low density because it has a lot

of empty space inside. Imagine, for example, a pile of large boulders tossed on top of each other. The boulders might have a density of, say, 3 g/cm³, but they are not packed into nice brick-and-mortar patterns; they are jumbled and there are many gaps – "void space" or "pore" space – between them. The density of this void space is effectively 0 g/cm³ and so the overall density of the whole pile is some number less than 3 g/cm³. It may be a lot less, depending on how loosely packed the boulders are.

What if that overall density turned out to be something like 1.5 g/cm³? Without having any other information except the density, then, we might interpret our pile of loose-packed boulders as a pile of ice instead, with a little rock mixed in. Clearly, we need additional information to try to determine whether the density is representative of a loose collection of debris, or a solid mass of material.

The first step in determining the density of Eros was

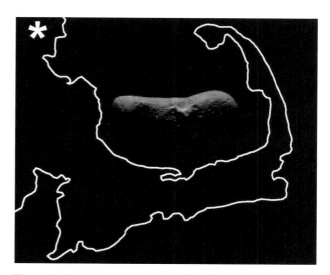

Figure 5.1. Eros shown on the same scale as the shoreline of Cape Cod, on the coast of Massachusetts. The city of Boston is indicated by the asterisk.

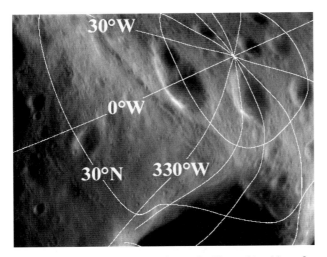

Figure 5.2. Latitude and longitude form a familiar and intuitive reference system on a spherical planet like the Earth, but not so on a small, irregular object like Eros. This NEAR image of the asteroid's north polar region was taken on March 31, 2000, from an orbital altitude of 207 km (129 miles). The image has been overlain with lines of latitude and longitude. The wandering, curved shapes of the lines are caused by the highly nonspherical and irregular shape of Eros.

to measure its mass. The mass of Eros was obtained by tracking the spacecraft: long-established principles from Newton's law of gravity and Kepler's laws of planetary motion allow the size and period of NEAR Shoemaker's orbit to be used to calculate the mass contained within Eros. Tracking the spacecraft involved combining radio measurements of its velocity with the imaging of particular landmarks on the surface over and over again (see Chapter 8). The NEAR Shoemaker spacecraft usually traveled at a speed of only a few miles per hour, and the distance to the center of Eros could be measured to within 100 m. The mass of Eros works out to about 6.69×10^{18} grams, or about seven million million metric tons.

Shaping up

The second step in the process was to measure the volume of the asteroid. For this a digital representation of its overall shape was required. Such a representation, or model, was also needed to make maps for use in locating the latitudes and longitudes of images, spectra, or other data from the mission. In the case of objects such as Earth, which can be represented as simple, mathematical figures such as spheres or ellipsoids, this is not a complex task. But for a pockmarked, banana-shaped object like Eros, it presents special challenges. Data from two instruments on NEAR

Shoemaker were used to deduce the size and shape of Eros: images taken by MSI and topographic data taken by the Near Laser Rangefinder (NLR). Both methods of determining the shape required independent knowledge of where the spacecraft was in relation to the asteroid, so combinations of radio tracking and following specific "landmarks" on Eros could be used to find the spacecraft's position relative to the asteroid (for more details see Chapter 8). For the image-based model, the position and scale of features in images were known; for the laser one, the distance to points on the surface could be subtracted from the spacecraft's position to yield a model for the shape of the surface.

The model developed by imaging depends on a process known as stereogrammetry, or the measurement of positions by using views from two or more directions. This is simply the quantitative application of what gives humans depth perception. The spacecraft's repeated orbits of Eros (more than one a day in the lower orbits) and the fast rotation of the asteroid (every 5.27 hours) combined to give images of the entire surface from a wide variety of directions. Points on the surface could be measured several times, allowing their relative positions to be pinpointed to within a few meters. This technique depended upon carefully marking the rims of craters and the pointed parts of boulders, such that they were consistently located to within fractions of a pixel regardless of the direction of view or lighting. The overall mathematical solution of the three-dimensional coordinates of all the points included over 30 000 individual measurements made on computer screens. Several people did this work over a period of months, adding points as more data arrived, and, in particular, as the Sun illuminated progressively more of Eros's southern hemisphere.

The marking required one to identify the same feature in different images and then accurately delineate its boundary. For craters this often meant marking six or eight points on the rim. The center of the crater is used in the triangulation solutions, so more points on the rim helped get better measurements. Before the shape model had been determined very accurately, it was often difficult to recognize features viewed from different directions because the predictions of where a previously marked point would appear in a new image were highly uncertain. Automatic recognition of the points was very difficult when the lighting had changed between one image and the next, so the human factor remained crucial. The small average errors of the measurements – less

Figure 5.3. Marking stereo points on Eros. These two views of the interior of Psyche crater were obtained from 50-km altitude orbit in May 2000 and show mostly the same area but from different directions. Craters that have been marked for the shape data base are displayed as colored ellipses. This figure illustrates how different some features look from moderately different viewing directions, as well as the problem of shadows hiding some points one might want to measure in an image.

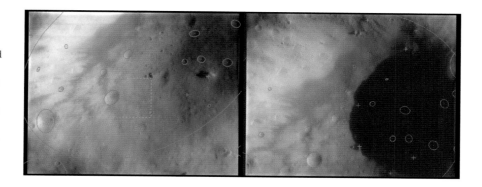

Figure 5.4. The Eros shape model draped over the limb of Eros. This view from 50 km shows how well the shape model conforms to curves on the asteroid's surface. The shape model is displayed as a network of triangular facets, derived from control points calculated using stereo information (see Figure 5.3).

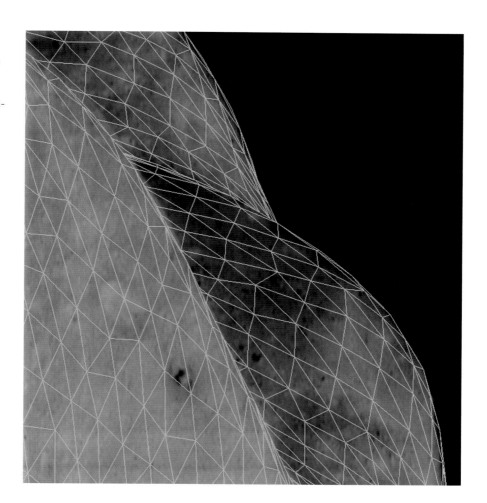

than 10 m – gave confidence that the model was a good representation of the real shape of the asteroid.

Having obtained the shape and size, we calculated the volume of Eros, which is approximately 2.53×10^{18} cm^3, or slightly more than the volume of water in lake Ontario on the US/Canadian border (which has a much greater surface area than Eros, but is very shallow compared to the asteroid's width). To get the density, we divided the mass by the volume and arrived at

2.64 g/cm^3. The density then has to be compared with values for likely materials – most importantly, meteorites.

A case of fracture

Establishing the relationship between the composition of Eros and that of the known meteorite types was one of the major goals of the NEAR mission, and was done

Figure 5.5. Jonathan Joseph, the Cornell NEAR team associate who wrote the software to derive the digital shape model of Eros from the NEAR MSI images, holds a physical model of Eros at 1:160 000 scale. The digital model was first converted to a laminated paper model, which was then the basis for a mold to make multiple copies with a resin cast. Having a physical model in hand made it much easier for scientists and mission planners to visualize many of the complex observational and orbital maneuvers executed by the NEAR Shoemaker spacecraft during the mission.

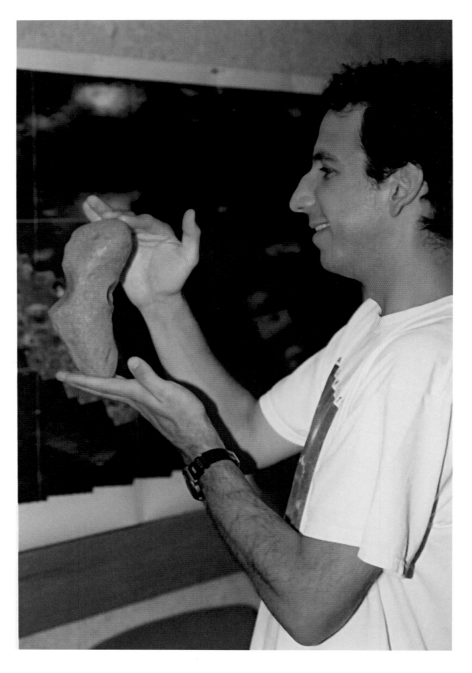

using data from the X-ray spectrometer (XRS) with additional constraints from NIS. The results from those instruments (see also Chapter 6) suggest strongly that Eros is made of materials similar to the so-called "ordinary chondrite" meteorites, which are a relatively primitive and highly abundant type of space rock. Densities of ordinary chondrite meteorite samples are typically in the range of 3–4 g/cm^3, with an average around 3.4 g/cm^3. The difference between this average meteorite density and the measured density of Eros may indicate the presence of voids in the rocky interior of Eros, created by fractures resulting from

impacts. The meteorites, of course, have also been subject to impacts, so the comparison is not perfectly straightforward. But the numbers suggest that there is about 20–30% void space caused by fracturing within Eros. One interpretation of NEAR's measurement of the density of Eros, then, is that the asteroid appears to have been heavily fractured, but not completely disaggregated.

But can additional details, such as variations from place to place in the porosity or in the composition of the interior of the asteroid, be inferred from NEAR data? Perfectly homogeneous rocks, miles long, are

Figure 5.6. Impact craters on Eros attest to the fact that it has been repeatedly bombarded, and that at least the layer near the surface of the asteroid is likely to have been heavily fractured by this rain of cosmic debris. This image, taken on June 10, 2000, from an orbital altitude of 51 km (32 miles), caught an obliquely illuminated view of a double crater. The two craters are so close to each other that they merge into the single dumbell-shaped depression in the center of the image. Each of the two craters is about 550 m (1800 feet) across. The whole scene is 1.9 km (1.2 miles) across.

very rare on Earth, so it would be surprising if this asteroid were exactly uniform throughout. The first test of the uniformity or homogeneity of the interior is to find out if the center of mass of the asteroid (the balance point, if we could magically place Eros on a see-saw) is at the same place as the geometric center (the center of the shape) of the asteroid. For a perfectly homogeneous object, these two centers will coincide. However, if there are lumps or large voids or different pieces of different compositions (rocks, metals) stuck together to build Eros, then these two centers could be offset from each other. This is the situation for example with "loaded" dice, which have a regular shape but an uneven distribution of mass within. Again, the tracking of the spacecraft, and accurate measurement of the shape from the images and laser altimeter, were used to locate these two centers. The result was that the asteroid's center of mass, the center of NEAR Shoemaker's orbit, is offset by between 35 and 50 m from the center of its shape (what would be the center of mass if it were homogeneous). This offset, a third of 1% of the average radius of Eros, is small, but requires some parts of the asteroid to be slightly more dense than others; Eros cannot be perfectly homogeneous. There is no single unique distribution of mass within Eros that could match this result, but very modest changes in pore space across the asteroid

could do it. Alternatively, it could be explained if the loose surface material – the regolith – were 100–200 m deeper on one end than another. But the images of Eros's surface materials and landforms do not support this latter explanation.

A matter of gravity

More can be discovered from Eros's gravity than just the total mass of the asteroid or the location of its center of mass. When the spacecraft was within 50 km of the asteroid the gravity field could be mapped by following the orbit of NEAR Shoemaker in detail. The slightly asymmetrical shape of the gravity field determined in this way was compared with what it would be if Eros were completely homogeneous in density. The two shapes are very similar, meaning that there are no large, dense sections, or particularly fluffy parts. The slight mismatch, as with the center of mass, could arise because of slight variations in porosity, or average chemical composition. But only very small ones are allowed by the close match of the physical shape and the gravity field. These findings are in concert with the NIS spectral results that show a fairly homogeneous surface.

Ridges and grooves

Another clue to the inside is to look for what is "sticking out" or, as in terrestrial rock outcrops, what structures are visible. Here the multiple views taken by the camera at different resolutions and under different lighting conditions were crucial in helping determine what is really a structure on the asteroid, and what is an artifact of changing illumination. The surface of Eros displays many different elongated features, and the trick is to try to interpret what they represent at depth. An 18-km-long ridge, Rahe Dorsum, is unusual for an asteroid. It probably represents slippage and breaking of rocks along the surface and some interior movement along a fault plane. Parts of this ridge are being compressed near the surface, and parts are pulling apart under tension near the ends. Because Rahe Dorsum makes an arc nearly halfway around the narrow part of the asteroid, it may not be just on the surface but part of a feature that cuts a few kilometers deep. Rahe Dorsum is surprising, but that surprise illustrates a well-known aspect of exploration: one is always trying to fit observations to previous experience. Nearly ten years earlier, on the first spacecraft flyby of an asteroid, Galileo's encounter with the

Figure 5.7. Views of the digital shape model of Eros, color coded by relative heights. The heights are those above what would be an arbitrary Eros "sea level," not from the center of the object. Red is high, blue is low. What is high and low depends partly on the spin of the asteroid (because of centrifugal forces), so it is not always intuitively obvious just what "relative heights" really mean on such an irregularly shaped object.

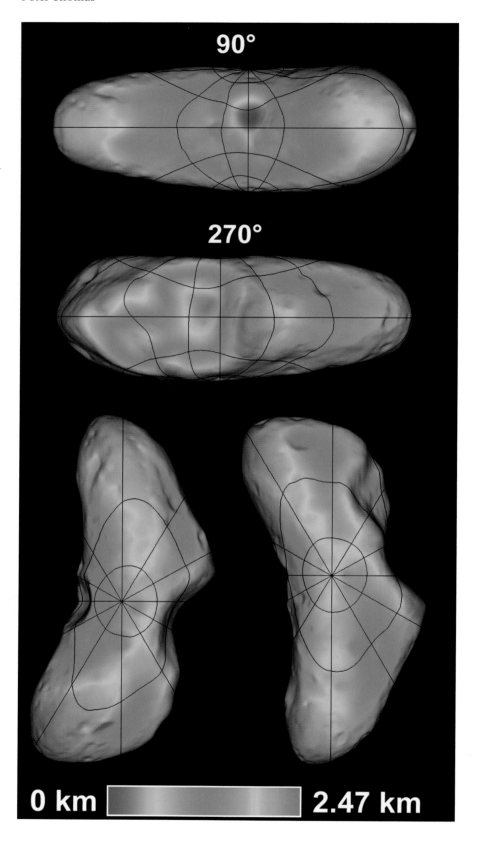

Figure 5.8. The low gravity (averaging only about 1/2000th the gravity on Earth, but varying by nearly a factor of two between different places on the asteroid) and the irregular shape of Eros often yield apparent optical illusions in NEAR images. For example, in this image, taken on July 22, 2000 from an orbital altitude of 43 km (27 miles), one of the asteroid's boulder fields decorates the skyline, but the orientation of the camera at the time the image was taken gives the illusion that the relatively gentle topography of the area is really a steep cliff. The whole scene is about 1.3 km (0.8 miles) across.

asteroid Gaspra, one image caught what looked like a short ridge running through a crater near the edge of the illuminated part of the asteroid. Although mentioned in publications, no one really dealt with it in overall interpretations because it did not conform well to expectations (and was poorly seen). Rahe Dorsum, however, is really obvious and will strongly influence interpretations of the mechanical properties of asteroids as scientists attempt to explain the origin of this structure.

Most linear markings seen on Eros are groove-like features, tens to a few hundred meters across, up to about 100 m in depth, and up to a few kilometers in length. Are these features layers, or fractures, or something else? Their "soft" morphology and commonly pitted appearance suggest that the narrower of them may be the response of loose materials above open fractures in the rock below. Most likely this would involve drainage of the loose regolith into open cracks

in the more solid material at depth, like sand through an hourglass. In this case though, the limited volume of the open cracks would allow only a fraction of the loose material to drain down, leaving small surface depressions following the fractures.

Drainage of material into cracks reminds us that, slight as the gravity is (less than 1/2000th of the surface gravity on Earth), it is effective in modifying the surface appearance of Eros. First, and foremost, the loose debris from the formation of impact craters that litters the surface, notably boulders and finer deposits, shows that gravity controls the flight of material ejected from impact craters. While it may be somewhat non-intuitive that small objects with such low gravity could retain ejected material, this same sort of loose regolith is also seen on many small planetary satellites, and on the small asteroids Gaspra and Ida (imaged by the Galileo spacecraft in the early 1990s). It is also predicted by the computer codes that simulate impacts of asteroids into one another. Secondly, gravity causes the loose material to move down slopes to the relatively lower areas. This is shown by streamers of material on crater walls and by the large number of ejecta blocks found in the bottom of some craters. These slopes, as on Earth, are simply those areas where the direction of gravity is not perpendicular to the surface. They are mostly fairly gentle on Eros. From the gravity measured at NEAR Shoemaker's orbit, and the shape of the asteroid, it is easy to calculate the angle between gravity and the surface. In most places these angles are less than 15°. Only over a small percentage of the surface are they above 35°. This is approximately the steepness that rocks and other loose debris assume when dumped as a pile, the so-called "angle of repose". Slopes substantially steeper would indicate coherent rock (though on Eros it does not have to be very strong). The surface topography of Eros has been pretty well pounded into gentle topography, but some features, such as Rahe Dorsum, appear young and steep.

Do the patterns of ridges, grooves, or other geologic features that we see in the NEAR images allow us to map the interior? The nearly homogeneous makeup of Eros suggests that it is not a collection of different pieces. The most obvious way to confirm this would be to find layers or other hard structures running the length of the asteroid. As in many terrestrial situations, getting a view of solid rock through soils and other loose regolith materials (what geologists call "overburden") turns out to be difficult on Eros. This is also true for some of the other small objects that have

Figure 5.9. Structural features on Eros.
Left: Rahe Dorsum, with grooves (top
right) and the "twist" (bottom right).
The distinct ridge of Rahe Dorsum is
unusual for small bodies, and probably represents compression due to
impacts. The grooves appear to be
loose material draped over or draining
into fractures in more solid material.
The "twist" is an ancient set of
grooves and ridges, highly degraded
by impacts and possibly by material
thrown out of craters.

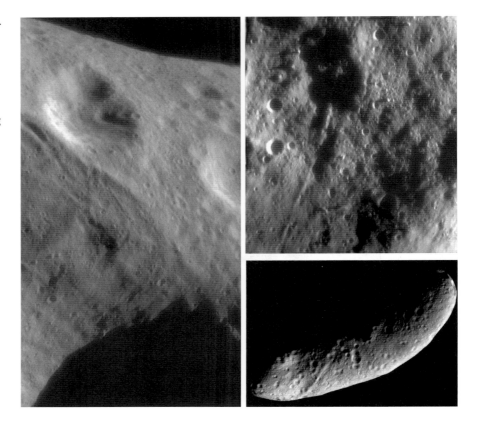

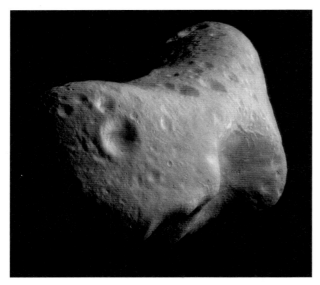

Figure 5.10. A mosaic of the northern hemisphere of Eros and the
large crater/depression Himeros (lower right) obtained by the MSI
instrument on March 2, 2000 from a 200-km orbital altitude.
Curving ridge systems within and around Himeros can be seen
in the lower-right part of the mosaic.

Figure 5.11. Ridges and other topographic features are enhanced in
oblique lighting conditions. Here, in this NEAR image obtained from
a range of 204 km (127 miles), a portion of the Rahe Dorsum ridge
system is prominently revealed. A series of closely spaced grooves
that follow the terrain downslope can also be seen along the left side
of the image. Features as small as 20 m (65 feet) are discernible in
this image.

Figure 5.12. Like sentinels in a lonely outpost, boulders and other loose debris often seem to accumulate in particular places within Eros's low-gravity surface environment. In this NEAR image obtained on May 18, 2000 from an orbital altitude of 50 km (31 miles), a field of worn and degraded craters is punctuated by a small gathering of jagged boulders, possibly formed when a single large ejecta block crashed back down on to the surface and broke apart. The angular boulder at the center of the frame is about 60 m (200 feet) tall, or two-thirds the length of a football field. The whole scene is about 1.4 km (0.8 miles) across, and it shows features as small as 4 m (13 feet).

covered with regolith, calls for patience and experience with views of a large variety of landforms on Earth and other planetary bodies.

A chip off an old block?

The result of a careful tabulation of surface forms that appear to reflect layers or fractures suggests that a global pattern of geologic structures (what geologists call a "fabric") exists throughout the asteroid, in addition to multiple local fractures caused by impacts. For example, Rahe Dorsum aligns with an older ridge and defines a planar surface cutting through most of the asteroid. Furthermore, part of the surface of Eros, just west of Shoemaker crater, is a fairly flat facet with an orientation similar to a plane defined by Rahe dorsum and the "twist" ridge described in Chapter 4. These alignments indicate that Eros retains some large-scale layers or fractures that it might have inherited from its larger parent body, and that have not subsequently been completely broken up and rearranged. The wide variety of other structures, though, shows that it has been severely fractured by non-catastrophic impacts.

In combination with the slightly non-homogeneous density of Eros, we may infer that the larger body assumed to have broken up to form Eros had subtle layering or mechanical fabric (fractures) enclosing areas of slightly different porosity and perhaps slightly different composition, on length scales of a few tens of kilometers. This structure may have resulted from the original accumulation of the asteroid, or from later, large impacts. From meteorites it is clear that the material forming the asteroids (and planets) accumulated from progressively larger pieces of material, initially the size of dust, but eventually many kilometers across. This gradual accumulation might generate layers of different materials. Layers and zones of

been studied up close by spacecraft missions (for example, Mars's moons Phobos and Deimos and the asteroids Ida and Gaspra). The surface is littered with debris from impact crater formation (see Chapter 4), and as there are few steep slopes on Eros it is not guaranteed that one could ever see nice layers exposed – in cliffs, for example. The search for global structures is thus a more subtle one. It requires the investigator to examine distant views under varying lighting as well as close-up views. Finding possible structures sticking out from beneath regolith, or structures that are still

Figure 5.13. A map of structural features (grooves and ridges), projected on to the shape model of Eros. Although there are many different directions of linear features shown, some have multiple examples, and individual craters do not show radiating fractures. These relations suggest that the grooves and ridges mostly follow very old fractures and other structures.

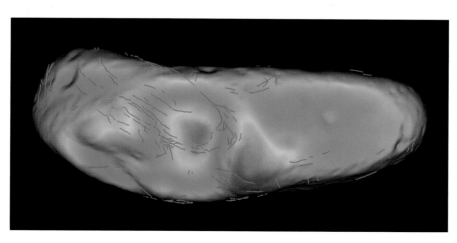

Figure 5.14. This mosaic of four frames, photographed on September 26, 2000, was taken as the NEAR Shoemaker spacecraft looked down on the Himeros "saddle" region from the south. The broad, curved depression that stretches vertically across the image is an area of the asteroid that was in shadow during the earlier 100-km orbit, in April 2000.

different materials might be caused by melting (there are many possible heating mechanisms, including energy from radioactive isotopes left over from supernova explosions near the solar system). Eros appears to have been broken off a larger object by a collision, so the question arises as to whether its structures relating to fractures arose before or after it broke off. Rahe Dorsum formed when Eros was essentially its present shape but, because of the properties noted above, probably shows some features that have been inherited from Eros's very early history.

The NEAR mission has shown us the great variety of features on the surface of a small asteroid and has given a partial glimpse of what is inside. This battered but mostly homogeneous rock is our current guideline for understanding much about the thousands of asteroids we see only as dots of light in ground-based telescopic images. Future missions may attempt to retrieve samples for return to Earth. These missions will probably first survey the asteroid for landing and sampling sites and thus may also give some context for the returned samples. To really determine what is inside an asteroid, landers with seismometers may be useful. Also, a lander that allows long-term radio tracking of the slight wobbles in an asteroid's rotation can give information on how rigid the object is. While there is a mission designed to expose some of the subsurface of a comet with a human-made impact (Deep Impact), no similar mission has yet been designed for an asteroid, though it might be most helpful in determining to what extent asteroids are truly "solid."

6 Ingredients of an Asteroid

Jim Bell

Cornell University (NEAR Science Team member)

Since Giuseppe Piazzi discovered the first asteroid, Ceres, in 1801 astronomers have catalogued more than 20 000 asteroids with orbits determined sufficiently well for their future positions to be predicted with a high degree of accuracy. And this is only a partial census, as perhaps 100 000 or more larger than 1 km in size still lurk in the vast spaces between the planets, yet to be discovered. Eros is one of approximately 1300 asteroids presently known to pass near Earth or to cross Earth's orbit. But despite the impressive body count, when added together, the masses of all the known or unknown asteroids probably do not amount to more than 10–20% of the mass of the Moon. For much of these last two centuries, asteroids were considered by many astronomers to be no more than "vermin of the skies," producing unwanted streaks or trails in long-exposure images of supposedly more worthy astronomical objects, like galaxies and star clusters. So why is there now so much interest in these insignificant and annoying bits of astronomical flotsam?

Asteroids are important for two main reasons. First, asteroids, and their close cousins the meteorites, are our most accessible surviving relics from the solar system's distant past. Many of the asteroids we see today are thought to be leftover building blocks that failed to form planets. They underwent most of their evolution during the first few hundred million years of solar system history and have since remained more or less as they were. In that regard, asteroids are tiny pieces of the early solar system, frozen in time. But they are not frozen in space, which leads to the second reason why asteroids are important: they sometimes collide with planets and their moons. Impacts have only recently been recognized as perhaps the most important process in the solar system responsible for modifying planetary surfaces. Asteroids are projectiles in an enormous game of cosmic billiards. One need look no further than Earth's Moon to appreciate the enormous devastation they have wreaked, especially billions of years ago when there were more of them around. And the Moon's well-preserved record of past impacts testifies that the Earth must also have been hit hard in the past. Indeed, more than 200 impact craters have been found on the Earth. Most are in a highly weathered and degraded state, but some, like Meteor Crater in Arizona, are still well preserved. In addition to the geologic havoc that impacts must have caused on the early Earth, there is now a consensus among astronomers, biologists, cosmochemists, and

Figure 6.1. The NEAR mission was designed to address fundamental questions in solar system studies, including the relationships between asteroids like Eros (center), and other similarly primitive bodies, and meteorites (right), which are presumed to be direct samples of some asteroids. Another question is the role near-Earth objects have played in modifying the geology of the planets and, in the case of Earth, the biology as well, through catastrophic impacts.

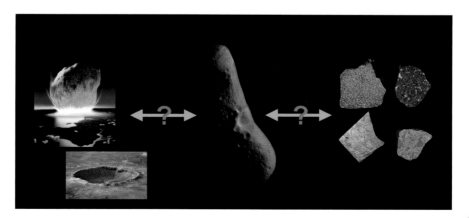

paleontologists that these events must also have caused biological mayhem, and were probably responsible for the relatively rapid mass extinctions of living species that have been identified in the fossil records. Are asteroids nothing more than insignificant cosmic leftovers? Ask a dinosaur.

But to truly learn about the asteroids requires that we know more than just their numbers and their orbits. We need to study asteroids from nearby. Only by close scrutiny can we learn about their origin and evolution and what their history can tell us of the formation of the solar system as a whole. The more we know about them, the better placed we are to assess the risks associated with impacts on Earth and possible mitigation strategies if we find that our planet is threatened. We have to find out what they are made of and how they are put together. Are they rocky, icy, or metallic? Are they solid and uniform throughout, porous and rubbly, or segregated into a core, mantle, and crust like planets? What rocks and minerals are found on asteroids? How did these materials form and under what conditions? Is the surface being modified by geologic processes and the space environment today? These are some of the most important questions that planetary scientists have been asking about asteroids, and they are also some of the most important questions being addressed by the NEAR mission to Eros.

Asteroid colors

One of the earliest ways that astronomers inferred the surface compositions of asteroids was to look at their colors. Specific rocks and minerals absorb or reflect sunlight of different colors in distinct, characteristic ways. Walking along a dry river bed or driving past a road cut, for example, it is easy to distinguish whitish silicon-rich minerals, such as sandstones, from reddish minerals with a high iron content, such as hematite. However, our eyes are tuned to only one small segment of the electromagnetic spectrum, what we call visible light, spanning wavelengths from about 450 nm (blue) to 650 nm (red). Many rocks and minerals also exhibit distinct "colors" at ultraviolet wavelengths (less than 400 nm) and in the near-infrared (wavelengths greater than 700 nm), both of which are invisible to human eyesight.

Astronomers often use a standard set of five color filters to observe objects through telescopes. This standard set is abbreviated to "UBVRI" for Ultraviolet, Blue, Visible, Red, and Infrared, and covers wavelengths from about 300 nm to 1000 nm. The first surveys of asteroid colors in the 1960s and 1970s used these filters to discriminate several different classes of asteroids and to develop a *taxonomy* of asteroid types, much like botanists have developed a taxonomy for plant species. The two most distinct classes are the

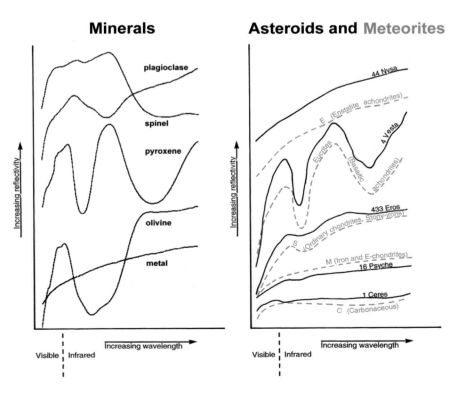

Figure 6.2. Spectroscopy provides a way to detect the presence of certain planetary surface minerals remotely. At left is a plot showing examples of spectra of minerals typical of asteroids and meteorites. Human vision is sensitive to wavelengths only near the leftmost part of this plot; instruments like NIS on NEAR are sensitive into the short-wave infrared as well. At right is a similar plot of telescopic spectra of specific asteroids (black lines, including Eros), compared to laboratory spectra of specific meteorites (red lines). The match between the spectra of certain asteroid classes and certain meteorite classes has provided evidence that meteorites are samples of asteroids, and that it is sometimes possible to link specific meteorites or meteorite classes with specific asteroids or asteroid classes.

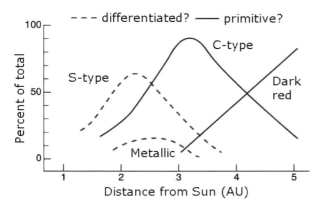

Figure 6.3. Asteroids are not randomly distributed throughout the solar system. Most are located in the main asteroid belt between the orbits of Mars and Jupiter. This belt appears to be stratified with different classes of asteroids predominating at different distances from the Sun, as shown in this schematic plot. The S-type, or silicate-rich asteroids (like Eros) include primitive objects as well as some that may have been partially differentiated because of the higher temperatures prevailing when they formed closer to the young Sun. The C-type, or carbonaceous asteroids (like Mathilde) are thought to be among the most primitive objects in the solar system, changed little from their initial formation more than 4.5 billion years ago.

Figure 6.4. The Galileo spacecraft obtained color images of main-belt asteroids 951 Gaspra (top) and 243 Ida and its small moon Dactyl (bottom) as it flew past these bodies on its way to Jupiter. Both objects revealed evidence for subtle variations in visible and near-infrared colors that were found to be primarily associated with impact craters. Apparently, on these two S-type asteroids, impact events have either excavated materials different from those on the surfaces of surrounding plains (suggesting that these asteroids are not homogeneous bodies), or have exposed similar materials that have different color properties because they have not been subjected to the "space-weathering" environment for as long as the surrounding plains. Gaspra and Ida are not shown to scale here.

C-type asteroids, which are dark and gray, and the S-type asteroids, which are moderately bright and reddish in color. The NEAR mission target, 433 Eros, is an example of an S-type asteroid, and 253 Mathilde, which NEAR Shoemaker flew past in 1997, is an example of a C-type asteroid. The early telescopic color studies revealed that the asteroids fall into different groups in different parts of the solar system. For example, most S-type asteroids are located near the inner edge of the main asteroid belt, while most C-type asteroids are located in the middle to outer parts of the belt. This distribution of asteroid types gives us information about the initial temperature, pressure, and compositional conditions of different parts of the solar system.

Colors can also be used to discriminate among different kinds of rocks and minerals on the surfaces of individual asteroids. Telescopic observers can infer the presence of different surface minerals by measuring an asteroid's colors, and can search for variations in the mineralogy between different places on the surface by watching the colors change as the asteroid spins. For example, an asteroid with one hemisphere rich in the mineral olivine (a common iron-bearing silicate on planetary and asteroid surfaces) and another hemisphere rich in the mineral pyroxene (another common iron-bearing silicate that forms under slightly different environmental conditions) would exhibit regular

variations if observed through a set of astronomical color filters as it spins. A possible conclusion would be that such an asteroid consists of fragments assembled from the debris of a shattered parent body. Alternatively, it might be an object with one or more large and deeply excavated impact craters on its surface.

Cameras on spacecraft missions can provide us with much more detail about the colors of planets, moons, and asteroids. For example, the Galileo spacecraft, which flew by asteroids 953 Gaspra and 243 Ida in the 1990s, used a camera with color filters that let through radiation in various bands from the ultraviolet to the near-infrared. Galileo's camera was able to record differences in color between different regions of these asteroids, and found that the largest variations in color, and by inference in surface composition, were associated with impact craters and their ejecta blankets. Apparently, on those objects, "fresher" materials have been excavated and exposed on the surface, producing areas of color distinct from the older, more "mature" background. NEAR's color camera has provided similarly detailed information on subtle color variations on the surfaces of Mathilde and Eros.

Spectra

Astronomers can obtain more valuable information about the colors of asteroids by means of spectroscopy. A spectrum displays how much light is reflected over a whole range of narrowly spaced wavelengths. Any dips in brightness at specific wavelengths that are known to be preferentially absorbed by certain kinds of minerals can be identified. One practical problem encountered with astronomical spectroscopy is that breaking light down into more component colors means that the signal at any particular wavelength is weaker and more noisy. Early spectrometers used for asteroid studies obtained data in 24 colors between 330 and 1100 nm or 52 colors between 800 and 2500 nm. More recent observations have measured the intensity at hundreds or even thousands of colors simultaneously over these wavelength ranges, producing high-resolution spectra that can be used to identify surface minerals. The broad, bell-shaped dips of the absorption features seen in these spectra are typical of asteroids. They are characteristic of absorption of light by the crystal lattices which make up all minerals. Gases, in contrast, usually exhibit very narrow absorption lines.

Spectroscopy measurements have also been made on a large number of meteorites. Meteorites are thought to be samples of asteroids, an idea strengthened by similar kinds of mineral absorption features in the spectra of a number of asteroids and meteorites. Specifically, the most primitive meteorites have spectral properties similar to the primitive, carbon-rich C-type asteroids. The spectra of several anomalous and apparently highly processed meteorites appear to match the spectra of some large asteroids that may once have been geologically active (see Figure 6.2). And, most interestingly for NEAR and studies of Eros, the spectra of typical S-type asteroids, which are essentially stony and either primitive or only slightly processed, have features in common with the spectra of the most common type of meteorite found on Earth, the ordinary chondrite class. The match between the S-types and the ordinary chondrites is not perfect, however, and so there is a question mark over the relationship between the two. Are S-type

asteroids such as Eros the parent bodies of the most common meteorites that fall to Earth? If they are, these rocks can be linked to a specific class of objects and a specific set of planetary formation conditions in the early solar system. If they are not, ordinary chondrites may only be samples of a small number of perhaps anomalous asteroids that happen to have ejected materials that found their way to Earth. In that case, they would be a biased sample, and basing general conclusions about the early solar system on them would be misleading. One of the most important goals of the NEAR mission was to try to determine if there really is a direct link between S-type asteroids like Eros and the ordinary chondrite meteorites.

An advantage of the laboratory study of meteorites compared to the remote sensing of asteroids is that many more kinds of measurements can be made on meteorites. An important example is the determination of the quantities of various chemical elements in meteorites by means of X-ray fluorescence or other techniques. These measurements produce the relative and absolute abundances of iron, silicon, oxygen, titanium, and other common rock-forming elements. Ratios of different pairs of elements have been used to classify the meteorites into different types and they also provide information on the geologic and/or environmental conditions on their parent asteroids. For example, the relative abundance of the highly volatile element sulfur in meteorites provides a simple way to discriminate between primitive, undifferentiated meteorite parent bodies and more highly evolved ones that were once more geologically active. Being able to make similar elemental abundance measurements of an actual asteroid promised enormous progress in identifying where meteorites come from. Results from the asteroid could be compared directly with those from meteorites and the data combined with the inferences from visible to near-infrared spectroscopy to help identify more precisely what minerals are present. And not surprisingly, this sort of synthesis was a major goal of the NEAR mission to Eros.

Instruments on NEAR Shoemaker

Four of the NEAR spacecraft's instruments – the multispectral imager, the near-infrared spectrometer, the X-ray/gamma-ray spectrometer, and the magnetometer – directly addressed the question of the composition and mineralogy of Eros. The radio science investigation provided additional indirect observations.

Figure 6.5. NEAR Shoemaker carried an array of scientific instruments designed to study a wide range of the electromagnetic spectrum, from high-energy gamma rays and X-rays, which probed the composition of the surface and subsurface, through low-energy radio signals that provided an accurate determination of the mass and gravity field of the asteroid.

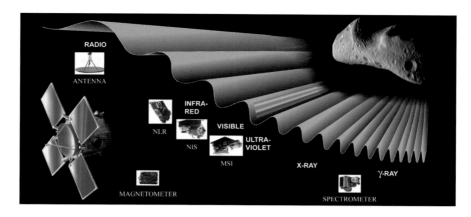

The multispectral imager (MSI) is a digital camera with an eight-color filter wheel. The eight colors were chosen to collect information on the overall color of Eros and on variations in the strength of mineralogical absorption bands across the asteroid's surface. MSI's strong point was not color coverage but spatial resolution – the ability to image features only a few meters across during orbits at the lowest altitudes. These images are very valuable for studying geologic processes on this small body. However, MSI's limited but well-thought-out color capabilities made it possible to search for variations in color that may be linked to composition and/or mineralogy and in turn associated with surface processes.

The near-infrared spectrometer (NIS) measured the spectrum of sunlight reflected from Eros in up to 64 colors between 800 and 2500 nm. NIS spectra can be used to identify the presence of a number of minerals seen in previous asteroid or meteorite spectra, and, more importantly, to map their distribution across the surface of the asteroid on a scale of several hundred meters. This spatial resolution is much coarser than that of the MSI but it is adequate for trying to relate potential variations in mineralogy as determined by NIS with geologic features, such as craters and ridges seen in MSI images.

The X-ray and gamma-ray spectrometers (XGRS) measured the spectrum of high-energy particles that are emitted by the surface of Eros when it is bombarded by protons and electrons in the solar wind. The energies of the emitted particles reveal the elemental composition of the asteroid's surface; that is, the total abundances of different chemical elements regardless of the minerals in which they may reside. Similar instruments flown on the Apollo lunar orbital missions provided important new information on the Moon's elemental composition. NEAR Shoemaker's XGRS instrument represents the first time the elemental composition of an asteroid was determined by remote sensing methods. This data set is crucial to establishing whether there is a link between Eros (and other S-type asteroids in general) and any meteorites collected on Earth.

NEAR Shoemaker's magnetometer (MAG) instrument was designed to search for and measure the intensity of any magnetic field that the asteroid might have. These measurements could provide information on the relative abundances of certain metals or metallic minerals on the surface and in the interior of Eros, thus providing information on the interior mass distribution and possibly on the asteroid's degree of differentiation.

The one other science experiment on NEAR Shoemaker providing insight into the composition of Eros was the Radio Science (RS) investigation, which allowed measurements of the mass and gravity field of Eros to be made. They were done by monitoring the change in NEAR Shoemaker's radio transmitter frequency as the spacecraft flew past and orbited the asteroid and comparing the results with predictions from computer models. Combined with the results on the shape and volume of Eros from MSI, RS gives us a measurement of the bulk density of Eros. The density tells us something about the interior composition – whether it is rock, metal, or ice, for example – and the gravity field provides clues to the asteroid's interior structure, which might be homogeneous or differentiated, porous, or solid.

The colors of Eros

Measurements from MSI were obtained during the NEAR orbital mission to determine the brightness of Eros in eight different colors, at wavelengths from 450 to 1050 nm. By terrestrial standards, Eros is not a very colorful object. To the human eye, it appears almost as

Jim Bell

Figure 6.6. Color images were
obtained with the NEAR MSI instrument by obtaining separate black-
and-white images through different
filters; the images were then merged
later by computer into color composites. For example, to make a true color
picture, MSI actually takes three
images through its red, green, and
blue filters. On the ground, the images
are carefully overlaid by computer (or
"registered") and displayed on a TV
monitor or in a hardcopy print in red,
green, and blue colors. Those steps
were used to render this color image
of Eros, taken February 25, 2000, from
a range of 349 km (217 miles).

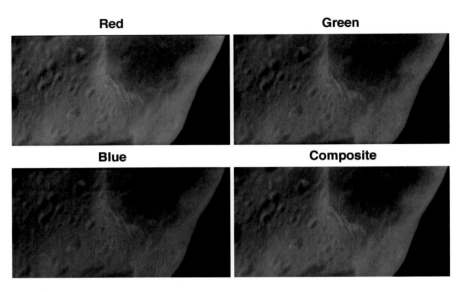

Red **Green**

Blue **Composite**

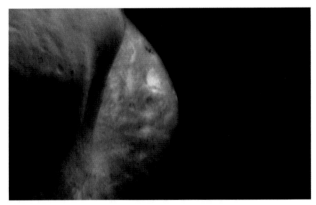

Figure 6.7. This color composite image of Himeros was taken on
April 2, 2000, from an orbital altitude of 201 km (125 miles). In this
false color representation, the red and green image planes were
taken in different wavelengths of infrared light, and the blue image
plane was taken in blue light. The bright and greenish-gray regions
near the rim of the saddle may represent relatively fresh (less "space
weathered") exposures of subsurface materials. In contrast, the
pinkish-looking surface materials covering other areas may have
been modified more heavily by exposure to small impacts and the
solar wind.

gray and colorless as does the surface of the Moon
viewed through a small telescope or as experienced by
the Apollo astronauts. Other asteroids studied up close
by spacecraft also show very little visible color on
their surfaces. But these other asteroids did show evidence for interesting ultraviolet or near-infrared
colors when they were studied with suitably sensitive
digital cameras. One of the major surprises of the
NEAR mission's digital imaging investigation was that
the brightness of Eros is extremely uniform throughout the spectrum, even at the ultraviolet and near-
infrared ends of the MSI camera's response range. No

color is much more prominent than any other. This
result is doubly surprising because some places on
Eros are more than twice as bright as others despite
the color differences between places being 10% or less.
Even in the highest-resolution color images obtained
by NEAR Shoemaker, there is no evidence that small
bright impact craters, where the subsurface might reasonably be expected to be exposed, are any different
in color from the rest of the surface. Color and brightness properties like these are to date unique among the
asteroids observed up close. By comparison, small
craters and other features show abundant evidence for
color variations on Gaspra and Ida, which are also
S-type asteroids, as well as on the Moon.

What could cause the appearance of high-contrast
bright and dark regions on the surface of Eros without
any significant variegation in the colors (and presumably, the compositions) of those regions? Scott
Murchie, a NEAR science team member from the Johns
Hopkins University Applied Physics Laboratory, and
other colleagues on the NEAR team have proposed the
hypothesis that the uppermost surface of Eros has been
modified by a process known for several decades in
asteroid studies as "space weathering." Space weathering has a similar effect on airless solar system bodies
to a process like rusting and oxidation on planets with
atmospheres (though of course the processes are different as Eros does not have an atmosphere). Over
time, interactions between the space environment and
the surface of an airless body alter the physical and
chemical properties of the surface, changing its color
and brightness. The ceaseless bombardment by a rain
of tiny impacting micrometeorites melts small quantities of the minerals in the uppermost surface layer,

Figure 6.8. Color composite view of the 5.3-km diameter crater Psyche obtained by NEAR Shoemaker's MSI camera on June 14, 2000 from an orbital altitude of 50 km. In this false color view, darker and redder hues represent rock and regolith that have been altered chemically by greater exposure to the solar wind and small impacts. Brighter and bluer shades, such as the bright patches on the crater walls, represent fresher, less-altered rock and regolith, possibly less affected by "space-weathering" processes.

allowing iron to be segregated out from the mineral structures and to coat other, unmelted, mineral grains. These sub-microscopic iron particles essentially act like a pigment, mostly darkening but also reddening the surface compared to its pre-weathered state. Murchie and colleagues propose that the space-weathered zone on Eros must be well mixed and quite thick – so thick that small impact craters do not poke through it. Large brightness differences are almost all confined to relatively steep crater walls, where even today tiny landslides and slumps can occur on Eros due to tidal forces or impact-induced seismic activity. When a small landslide does occur, underlying layers of materials that have been subjected to slightly less space weathering are exposed to view, and these materials appear brighter. Other researchers, including Bruce Hapke of the University of Pittsburgh and his co-workers, have speculated that space weathering is a relatively fast-acting process in the near-Earth environment. Astronomically speaking, fast-acting means

a noticeable effect over "only" tens to hundreds of millions of years. Leading on from this, we can reason that NEAR Shoemaker did not observe any color variations associated with the many large impact craters on Eros, which presumably could have excavated deeper regions of the subsurface, because those rarer substantial impacts occurred so long ago that the exhumed subsurface materials have themselves been space weathered. They share the relatively dark and colorless appearance of the rest of the asteroid. If these ideas are correct, then it would appear that Eros has not suffered a large, deeply excavating impact event within recent times in its geologic history.

Minerals on the surface

The NIS instrument on NEAR Shoemaker obtained near-infrared spectra of sunlight reflected off the surface of Eros from the time of initial orbit insertion on February 13, 2000 until the instrument failed unexpectedly on May 13, 2000. Despite the untimely end of the NIS investigation, spectra of nearly 90% of the surface of the asteroid were secured during the three months in which it was operational, including a number of critical observations from unique viewing directions never before achieved during a spacecraft encounter with an asteroid. NIS obtained more than 200 000 reflectance spectra of Eros's surface, resolving variations over distances as small as approximately 250 m, plus nearly 200 000 additional spectra during the cruise phase for instrument calibration and testing, as well as observations of the Earth and Moon, and of Eros during the December 1998 flyby.

The NIS spectra of Eros confirm the presence of the minerals olivine and pyroxene on its surface and indicate a ratio of olivine to pyroxene that is similar to the ratio observed in many ordinary chondrite meteorites. These findings are consistent with unresolved observations of Eros made by ground-based telescopes in the 1970s and 1980s, and validate the methods used by astronomers to interpret the spectra of the thousands of other asteroids that have been observed with the aid of telescopes but not visited by spacecraft. My own studies of the NIS data, with colleagues Paul Lucey (University of Hawaii), Beth Clark (Cornell University), and others indicate the possible presence of other minerals on the surface, in addition to olivine and pyroxene, including possibly iron sulfide, iron-nickel metal, a bright component such as plagioclase, and, possibly, relatively large amounts of impact-generated and relatively iron-rich glass.

Figure 6.9. Interpreting NIS spectra often involves computer modeling and comparisons with spectra of minerals, mineral mixtures, and meteorites. In (A) a typical NIS spectrum of Eros (open squares, bottom) is compared to the spectrum of a typical ordinary chondrite meteorite (crosses, top). This meteorite is brighter than Eros typically appears and has a somewhat different spectrum. The solid lines represent a model that attempts to explain the differences between Eros and this meteorite by adding small amounts of submicroscopic iron – like that generated by space-weathering processes – to the meteorite spectrum. The brightness can be made to match, but not the shape of the spectrum.

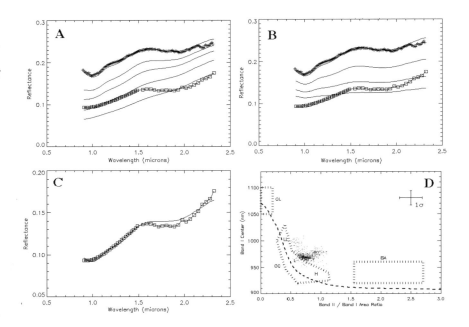

In (B) a similar attempt has been made, but in this case by adding a relatively gray darkening agent (like carbon or iron sulfide for example) to the meteorite spectrum. Again, the brightness can be made to match, but not the spectral shape. (C) represents the effect of combining these results, as well as allowing the silicates to have a relatively coarse grain size, in order to match both the overall brightness and the spectral shape. It is possible that darkening and reddening processes over long timescales have altered the spectrum of Eros from its "pristine" form, which may have been similar to that of ordinary chondrite meteorites.

Additional support comes from the plot in (D), based on the work of Michael Gaffey, Edward Cloutis, and colleagues. Analysis of the details of the absorption bands seen in the NIS spectra shows that Eros spectra (the cloud of data points in the lower left of this plot) appear to exhibit similarities to the properties of primitive, ordinary chondrite (OC) meteorites, rather than to the more-processed olivine-rich meteorites (OL) or basaltic meteorites (BA).

Because they were acquired so close to Eros, the NIS spectra can be used to zoom in on the spectral properties of small geologically interesting regions of the asteroid's surface. One of the most surprising results of the NIS investigation was that the spectra exhibit very little variability from region to region. Despite large differences in the geology of the surface (craters, ridges, boulders, etc.), Eros is remarkably homogeneous in its spectroscopic properties. The largest differences, which are only a few percent deviations from the average, are associated with the most extreme surface slope conditions, such as within steep crater walls. The inference is that Eros's composition and mineralogy are practically the same everywhere, and that the observed variations mostly reflect the influence of external, rather than internal forces throughout the asteroid's history. For example,

if Eros's interior were differentiated into layers of dissimilar density or composition, or if it had once been part of a larger differentiated parent body, then one might expect to see differences in composition and mineralogy from place to place, perhaps near craters that have dug up the subsurface, or near ridges where tectonic forces have modified the landscape. Spacecraft observations of the asteroids Gaspra and Ida, and telescopic measurements of other main-belt asteroids, have detected surface diversity of that kind. However, no such variations were identified on Eros.

It is possible that the space-weathering process has effectively "washed out" variations in surface mineralogy, so that all parts of the asteroid look pretty much the same regardless of the wavelength of the radiation used for the observations. If this is true, it supports the idea that Eros has not been hit by a major impactor recently. But is there really just a thin veneer of uniform material overlying a range of different compositions in the subsurface and interior of Eros, or is the asteroid truly homogeneous throughout? Interpretation of the NEAR Radio Science (RS) gravity and density measurements support the latter hypothesis. Specifically, the mass of Eros was determined by the RS during the initial flyby in 1998 and, when combined with the volume of the asteroid determined by MSI images, yields an estimated bulk density of approximately 2.67 grams per cubic centimeter. This density is within the typical range of ordinary chondrite meteorites, most of which are fairly uniform in

Figure 6.10. Planning and acquiring observations from NEAR Shoemaker's Near Infrared Spectrometer (NIS) was a complex process that required a detailed understanding of the shape and the rotational state of Eros. On February 13–14, 2000, just before insertion into Eros orbit, NEAR Shoemaker passed between Eros and the Sun. This relative positioning is called a low phase angle. It is optimum for spectroscopic observations, so at that time NIS obtained some of its most valuable data. To plan these activities in the preceding months, the NEAR science team developed a series of computer-simulated observation sequences like the one depicted in the left image. Here, NIS was instructed to scan its mirror back and forth across Eros while the spacecraft did a slow slew from top to bottom. The image shows the predicted NIS footprints as viewed from directly over the north pole of a computerized shape model of Eros.

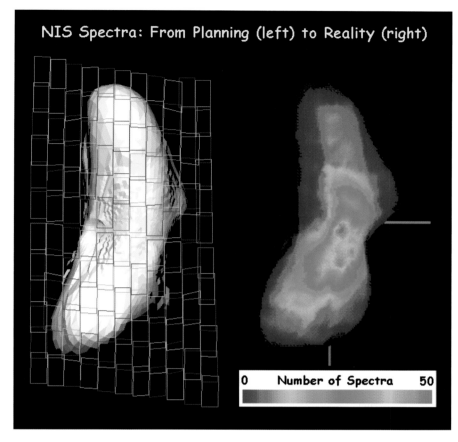

Many such observation sequences were performed flawlessly on the spacecraft during the low-phase flyby, and the spectra of highest quality were combined to show the resulting NIS coverage of Eros in the map on the right. The green and red regions show areas at high northern latitudes with the best coverage by NIS, with up to 50 spectra covering some regions near the north pole.

composition and mineralogy. It is not indicative of a particularly icy or metallic interior composition, which would result in a lower or higher density respectively. Perhaps more important, however, are the results from the RS investigation obtained while NEAR Shoemaker was orbiting Eros. During this time, careful analysis of small Doppler shifts in the spacecraft's radio signals were used to construct a map of the asteroid's gravity field, from which the presence and magnitude of internal density variations could be inferred. After compensating for the asteroid's irregular shape, which gives rise to a rather bizarre and non-intuitive surface gravity distribution, the gravitational field of Eros was found to be consistent with a remarkably homogeneous interior density. Eros appears to be the same throughout.

One additional set of NIS observations contributes to our understanding of the surface of Eros. Just before going into orbit, NEAR Shoemaker made a series of specially designed NIS observations taking

advantage of a unique opportunity to view Eros illuminated in a particular way. The spacecraft was programmed to pass Eros with the Sun almost exactly behind the NIS instrument's field of view, what astronomers would call observing at a 0° phase angle. This particular kind of lighting is highly favorable for spectroscopic observations because it minimizes the amount of shadowing on the surface. The nominal viewing geometry for most MSI images and other orbital measurements was a phase angle of 90°. This means that there is an angle of 90° between the line of sight from the instrument on the spacecraft and the direction from which sunlight is striking the asteroid's surface. By observing the same surface regions over a range of phase angles from 0° through 90°, it is possible to infer additional information about the composition and physical nature of the uppermost surface of the asteroid, such as the size of grains, large-scale roughness, and intrinsic brightness (albedo). By analyzing NIS spectra over a wide range of phase angles, NEAR scientists Beth Clark and her colleagues have been able to compare the physical characteristics of the surface of Eros with those of the other asteroids and small bodies observed by spacecraft. In general,

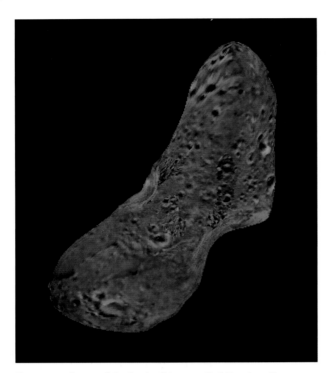

Figure 6.11. Image of the depth of the so-called "1-micron" (1000-nm) absorption band in NIS spectra of the northern hemisphere of Eros. A map of the strength of this absorption band has been draped over the computer shape model of Eros; craters and other features seen in MSI images are also shown for context. The average band depth is about 18% (green areas in this image), and the variations across the asteroid are extremely small, ranging from a minimum of about 16% (blue) to a maximum of about 19% (red). Further, the variations are confined to only a few small areas, like parts of the walls of Psyche and Himeros, and near the 0° longitude "nose" of the asteroid. This suggests that the surface overall is remarkably homogeneous in its physical and mineralogic properties, at least at the scale of 2–3 km represented by the NIS data shown here.

the surfaces of asteroids and similar bodies appear brighter when observed at smaller phase angles. In the case of Eros, the effect was less marked than for Gaspra or Ida. A possible interpretation is that Eros is composed of a mixture of materials that scatter light more efficiently at all angles than the materials found on Gaspra or Ida. One of the most useful applications of the special observations at low phase angles was the development of a system to compare NIS measurements made over a wide range of viewing and illumination conditions with laboratory measurements of rocks and minerals, typically made under a single fixed set-up. The ability to compare NIS data directly with laboratory data was essential if we were to identify the different minerals that could be responsible for the observed NIS spectra of Eros.

The basic elements

The final pieces of data to help solve the puzzle of Eros's composition came from the NEAR XGRS instrument. During several large solar flares that happened to occur during the mission, the XGRS instrument obtained X-ray energy spectra of regions on the surface of Eros. These yielded estimates of the relative abundances of iron, silicon, magnesium, calcium, and sulfur. Initial determinations of the Eros elemental abundances were directly compared to the elemental abundances of different classes of meteorites, and all elements except one were found to be generally consistent with the chemical composition of the class of meteorites called ordinary chondrites. While it is not possible to determine if some of the ordinary chondrites in the world's meteorite collections are actually pieces of Eros itself, the asteroid's general proximity to

Figure 6.12. A global map of the gravitational field of Eros as determined by analyses of the Radio Science investigation data by science team members Don Yeomans, Maria Zuber, and their colleagues. The measured gravity (upper left) is subtracted from the gravity expected from a simple homogeneous body having Eros's shape (upper right and lower left), and the residuals (lower right) provide information on any "lumpiness" or interior structures within the asteroid. To within a few percent, the interior of Eros appears to be quite homogeneous.

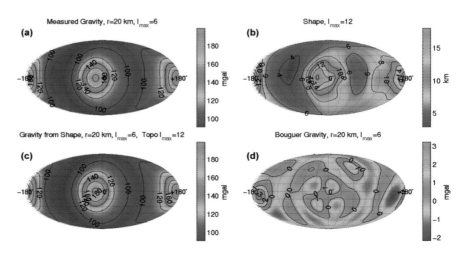

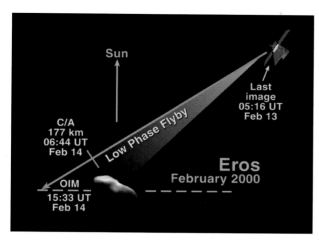

Figure 6.13. On February 13–14, 2000 the NEAR spacecraft flew directly between Eros and the Sun, in a maneuver called the low phase flyby (LPF). This geometry is optimal for spectroscopic measurements because it minimizes shadows on the surface. During the LPF, a number of complex spacecraft maneuvers had to be executed in order to obtain these critical NIS measurements. Perhaps the most complex and elegant was when the spacecraft had to match its spin with that of the asteroid while flying past, in what was termed the "2001 maneuver" – named because the relative motion of the spacecraft and asteroid was similar to that of the shuttle and the Earth-orbiting space-station in the famous rendezvous scene in the movie "2001: A Space Odyssey." All of these complex operations were executed flawlessly thanks to the hard work of engineers and scientists on the mission operations, navigation, and science sequencing teams.

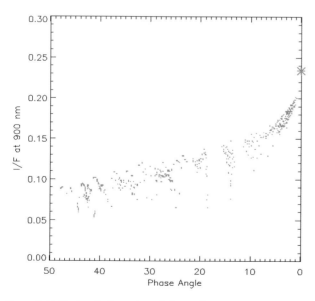

Figure 6.14. During the low-phase flyby, as the spacecraft got closer to passing through the Eros–Sun line (that is, as the phase angle got lower), the spectra measured by NIS got brighter and brighter. This well-known "opposition effect" had never before been studied in such detail for an asteroid, and the data provide important clues about the physical nature and composition of the surface of Eros.

Earth and occasional close passes to us support the idea that Eros may be the parent body of at least some of the ordinary chondrites. The elemental composition of Eros is clearly not consistent with the compositions of any of the highly evolved or differentiated meteorite classes – the achondrites. This indicates that Eros has likely retained its relatively primitive composition since its initial formation early in the history of the solar system. The one anomalous element was sulfur, which appeared in some of the XGRS data to be substantially less abundant on Eros than in almost any type of meteorite. This could be a measurement artifact or error: the XGRS team was continuing to refine their results as this was written. On the other hand, it could reflect the fact that sulfur is one of the most volatile elements. If Eros underwent even mild internal processing – perhaps in the initial phases of planetary differentiation – then the sulfur would have separated out and been lost first. Alternatively, there could be something that we do not yet comprehend about the space-weathering process. It may preferentially drive off sulfur compared to other elements. The observations of very low sulfur abundances in some of the XGRS spectra of Eros are an intriguing result, but one that is not yet well understood.

Science from the surface

Much to the delight of everyone on the science team, the XGRS and MAG measurements of Eros obtained from orbit were augmented by additional data obtained during and after NEAR Shoemaker's spectacular February 13, 2001, landing on the asteroid. No one had expected the spacecraft to survive the landing, let alone obtain any scientifically useful measurements, and so the magnetic field measurements during descent and the subsequent two weeks of gamma-ray measurements obtained on the surface represent a real scientific bonus for the project. The early analysis of the landed gamma-ray compositional results is generally consistent with the orbital elemental results – the surface appears to have a composition essentially similar to ordinary chondrite meteorites. However, iron appears to be somewhat depleted in the gamma-ray data relative to ordinary chondrites. This may indicate that iron-bearing minerals like iron sulfide may be segregated within Eros's regolith, at least at the landing site, or it could be a measurement artifact of some kind. The gamma-ray spectrometer within XGRS was never intended to make measurements while buried under the spacecraft within the

Jim Bell

Figure 6.15. During a solar flare on March 2, 2000 the NEAR XGRS instrument was able to measure X-rays emitted from the surface of Eros in response to the increased activity of the solar wind. The top plot shows highlighted in red on a shape model of Eros the region that was excited by the solar wind and measured by the XGRS instrument. The bottom plot shows the resulting X-ray spectrum. Peaks in energy can be seen that correspond to X-rays emitted from rock-forming elements like Mg, Al, Si, Ca, and Fe. The intensity of the peaks is proportional to the abundances of these elements on the asteroid's surface.

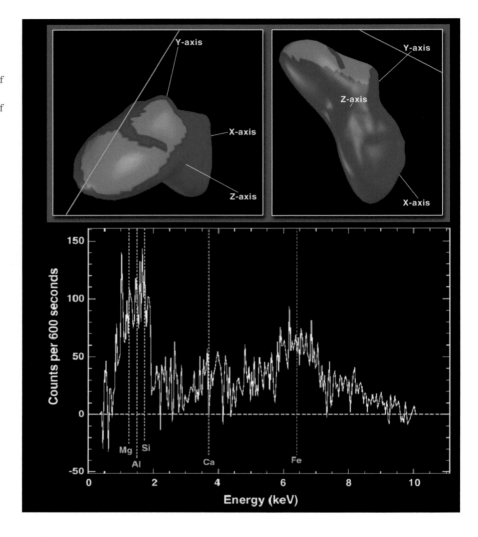

Figure 6.16. This plot shows the gamma-ray spectrum measured from the surface of Eros. These scientific data – the first ever-collected on the surface of an asteroid – result from seven days of measurements following NEAR Shoemaker's historic landing on February 12, 2001. The gamma-ray instrument has two detectors – marked above by the red and blue traces – which picked up clear signatures of key, rock-forming elements on the surface of Eros. These data, which surpass in quality all the data accumulated by this instrument from orbit, are helping NEAR scientists relate the composition of Eros to that of meteorites that fall to Earth.

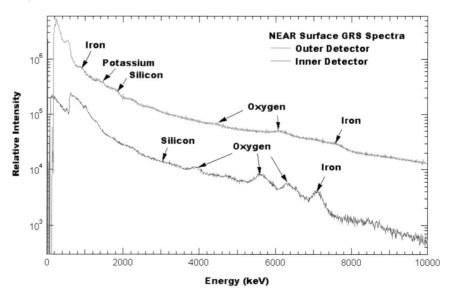

surface of Eros! The MAG measurements are also
interesting, but for what they *do not* reveal. No magnetic field was ever detected for Eros during the orbital
phase of the mission, and so it was hoped that perhaps
a weak intrinsic field could be detected directly on the
surface. However, not even a weak field was detected
in the measurements obtained by MAG during the
final descent or after the landing. This result is intriguing and not yet clearly reconciled with other measurements from the mission. Specifically, if Eros is
composed of the same materials as ordinary chondrites, then it might be expected to contain some
metallic, and thus likely magnetic, minerals on its
surface and in its subsurface. However, Eros appears
much less magnetic (if at all) than many of the types
of meteorites that it appears most closely linked with
chemically and mineralogically. It will be interesting
to see how these first scientific measurements and
results from the surface of an asteroid can be reconciled with the more extensive orbital measurements
after scientists have had time to perform additional
detailed analysis and interpretations.

NEAR's testimony

New and unique data from the NEAR mission are
enabling us to assess the composition and mineralogy
of Eros in a detailed way and to explore the relationships between meteorites and their parent bodies, the
asteroids. Because NEAR's discoveries about the composition of Eros's surface are broadly consistent with
earlier results obtained using telescopes on Earth,
astronomers can now have confidence that their telescopic techniques are capable of accurately assessing
the properties of other members of the enormous population of asteroids that cannot be observed close up
by space missions at the present time. More importantly for asteroid studies in general, NEAR's evidence
generally supports a direct link between one broad
class of asteroids – the S-types – and meteorites of the
ordinary chondrite class. While these interpretations
are not without controversy, and detailed analysis of
the data continues, NEAR appears to have revealed
that Eros is a relatively primitive remnant of the early
solar system, modified only slightly by external forces,
such as impacts, and by external physical and

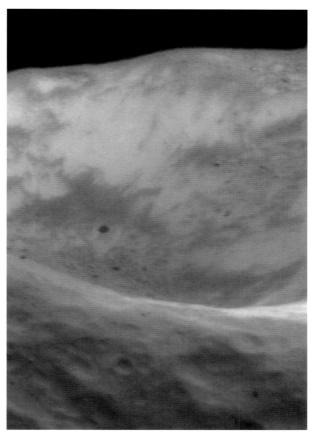

Figure 6.17. A high-resolution false-color image of the floor and
walls of the 5.5-km-diameter crater Psyche, based on the work of
Scott Murchie and colleagues. The colors depict variations in the
strength of the "1-micron" absorption feature on Eros at a spatial
scale of only 4 m (12 feet) per pixel – much better resolution than
could be obtained from the NIS instrument. Parts of the rim and
walls of Psyche exhibit a deeper absorption feature (brighter and
redder in this false color view), consistent with this material being
"fresher," or exposed to space-weathering processes for less time
than the surrounding materials. This is consistent with other
imaging evidence that shows that slumps, landslides, and other
mass-wasting processes have acted to mobilize parts of the regolith
of this small body, even in relatively recent geologic times.

chemical alteration processes like space weathering. If
the suggested link between Eros and the ordinary
chondrites holds up in the coming years while more
detailed analyses of the NEAR data are carried out,
new telescopic and spacecraft studies of asteroids may
finally reveal to astronomers the nature and composition of the early solar system.

7 The Battered History of Eros

Clark R. Chapman
Southwest Research Institute, Boulder, Colorado (NEAR Science Team member)

It was uncanny. It was Christmas Eve 1998. I stood in a small, computer-crowded room called the NEAR Science Data Center, within the labyrinth of the sprawling campus of the Johns Hopkins University's Applied Physics Laboratory (APL) in rural Laurel, Maryland. I marvelled at our first pictures of Eros, radioed back from the NEAR Shoemaker spacecraft, weeks before we were scheduled to get our first good look at the asteroid. Though the pictures were brand new, Eros looked strangely familiar. The 34-kilometer-long, Earth-approacher had the general shape of a kidney bean, as expected from the crude model of its shape based on radar echoes published years earlier. Eros's personality, as viewed from a range of about 4000 km (2500 miles), was dominated by an enormous gouge carved into its outer periphery. Such depressions are missed by radar, which detects only a shape devoid of concavities, as if a rubber sheet were stretched across and around the asteroid. A year later, NEAR Shoemaker's camera would map this gouge, since named Himeros, in much more comprehensive detail, revealing it to be a saddle-shaped depression as wide as Eros is thick. Later, after a year orbiting Eros, the spacecraft was directed to make a daring landing near the rim of this enormous topographical depression.

I kept feeling that I had somehow seen this gouged visage of Eros before. Someone handed me a copy of *Asteroids II*, the most definitive book on asteroids to date, published in 1989. There was Eros on the book's cover! Years before the NEAR mission was selected as one of the first of NASA's so-called Discovery missions – developed under the slogan of "Faster/Better/Cheaper" (FBC) – Bill Hartmann, a well-known space artist as well as a planetary scientist, painted his vision of what an Earth-approaching asteroid might look like. Except for being a mirror image, his painting *is* Eros! Hartmann's imagination had eerily conjured up

a perspective that matched not only its overall shape, including Himeros, but even some smaller features. Yet the stranger, and more portentous, situation on that Christmas Eve was that we had images of Eros at all, two-and-a-half weeks before the spacecraft's scheduled arrival.

My 1994 proposal to participate in the NEAR mission had focused on synthesizing data from four instruments concerning Eros's chemical and mineralogical composition. But each member of an FBC mission's pared-down science team must handle multiple scientific responsibilities, so I was also to study craters, as I had done using pictures taken during previous spacecraft flybys of the asteroids Gaspra, Ida, and Mathilde. Another responsibility, shared with my associate Bill Merline, was to find out whether Eros had any small moons orbiting around it. Asteroids had been thought to lack moons but that all changed in early 1994 when pictures from the Galileo mission to Jupiter showed a moonlet 1.6 km (1 mile) wide orbiting around the asteroid 243 Ida. As a result, asteroid moons had become a hot topic. The optimum time to search for any satellites would be during the weeks before the spacecraft reached Eros, when the camera would map the adjacent sky before finally homing in on the asteroid itself.

Because of this, two months before other science team members would take a serious interest in NEAR's early images of Eros, Bill Merline and I started examining the far-away views for signs of a small companion. In the wee hours of December 22, Bill was at his computer terminal in the Boulder, Colorado, offices of the Southwest Research Institute looking in vain for the latest image files of Eros. They were to have been snapped during the hours following the previous evening's "big burn," when NEAR Shoemaker's main engine was to have slowed down the spacecraft in preparation for orbital insertion on January 10. Bill

Figure 7.1. When the first images revealing the actual shape of Eros came back from the NEAR mission, they bore an amazing resemblance to a rendition of a near Earth asteroid painted in 1989 by planetary scientist and space artist Bill Hartmann for the cover of the book *Asteroids II* published by the University of Arizona Press. (This NEAR Shoemaker image is flipped left-to-right.)

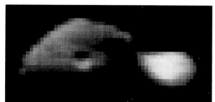

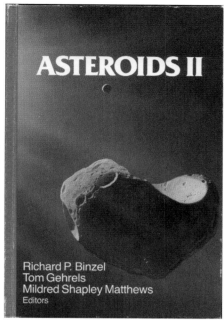

soon feared what would prove to be true: the burn had failed. Indeed, it had aborted after just two seconds, after which NEAR went into a wild tumble and all contact with the craft was lost.

Initially it seemed like a repeat of events in 1993, when the Mars Observer spacecraft abruptly disappeared, apparently having exploded, as it approached Mars. But, later in the day, contact with NEAR Shoemaker was re-established and the spacecraft seemed to be healthy. Frantic preparations began for making what observations we could when NEAR Shoemaker zoomed by Eros the very next day, 18 days before planned arrival. At Cornell University, Maureen Bell and Ann Harch worked around the clock to update an imaging sequence they had already designed for this kind of contingency, in the hope of capturing some pictures of Eros as NEAR Shoemaker flew past. The coded commands were radioed up to the spacecraft with only minutes to spare.

I had to get to APL. My wife, Lynda, had long planned to accompany me to APL when NEAR Shoemaker arrived at Eros, so she came along on my premature, emergency trip. We negotiated pre-Christmas crowds, a snowstorm that crippled mid-west airline hubs, and a slushy Washington Beltway in order to get from Colorado to Maryland. While we travelled, NEAR Shoemaker calmly executed its new instructions as it sailed past Eros, 375 million km (233 million miles) away across interplanetary space. Within hours of arriving at APL early the next morning, Bill and I studied the newly downloaded

images. We found no obvious moonlet. Now I was having my first glimpses of the crater-scarred surface of Eros. For all I knew, the half-dozen visible craters would be all that we would ever learn about this small world's geology. Fortunately of course, NEAR Shoemaker survived to try again and successfully entered orbit around Eros a year later.

Battered remnant worlds

The cosmic impacts that have shaped the surfaces of most solar system bodies are an inevitable aftermath of the origin of the planets. The terrestrial planets and the cores of the outer planets grew by the accumulation of a hierarchy of countless smaller bodies, termed planetesimals. The planetesimals themselves had grown from the solid materials condensed out of the solar nebula. As more and more planetesimals became incorporated into nearly fully grown planets, the supply of them diminished. Those that came close to the planets were diverted by the gravity of these larger bodies on to new paths. Some were despatched out of the planetary system, perhaps into the Oort comet cloud. Others were sent crashing into another planet or the Sun. But, in zones far from planets, reservoirs of planetesimals remained. The rocky asteroids settled in a belt between the widely spaced orbits of Mars and Jupiter and the predominantly icy Kuiper Belt objects accumulated far beyond Neptune's orbit.

Over the aeons, chaotic gravitational and orbital forces cause some asteroids and comets to "leak" from these reservoirs and follow elliptical orbits that cross the paths of major planets; some soon crash into the surfaces of moons and planets, pitting them with impact craters. Meanwhile, all of these small bodies collide with one another: their fragmentation creates yet more small projectiles and their own surfaces are struck and cratered as well. Impacts continue to pockmark the surfaces of planets, disturb their atmospheres, and – on Earth – shape the evolution of life. On worlds where active geological processes persist, such as Earth, most evidence of past impacts is erased. But collisions and cratering impacts dominate the post-formation history of the geologically inert asteroids.

The dynamical evolution and impact record of asteroids has special relevance to our own planet, as well as to the scientific study of processes in the early solar system. Impacts among asteroids create fragments, and chaotic dynamics divert such fragments on to planet-crossing and Earth-approaching orbits. On rare occasions, during their transient existence between the asteroid belt and crashing into the Sun (or ejection from the solar system), a near Earth asteroid (NEA) may strike our planet, perhaps with dire consequences for living things. Far more frequently, much smaller, fist- or automobile-sized fragments (meteoroids) are extracted from the asteroid belt and arrive on Earth's surface as meteorites. A small proportion of them are found and studied for clues about their asteroidal parent bodies.

In gleaning insights from meteorites about our planetary system's formative epochs, we have two obstacles to overcome: (1) identifying *where* a meteorite comes from, since its chaotic path to Earth makes it impossible to determine the address of its previous residence in the asteroid belt, and (2) disentangling the damage done by four billion years of impacts. A meteorite that has been on its parent body's surface for the full duration, will have been damaged beyond recognition. Fortunately, typical meteorites are protected in the depths of their parent bodies until liberated by a collision so they are less damaged (by impact shock and melting, for example) than most rocks collected from the regolith (the heavily impacted and fragmented upper layer) on the surface of the Moon.

Consider NEAR's target, Eros. Its current, temporary, planet-crossing orbit around the Sun stretches out beyond the orbit of Mars, practically to the inner edge of the asteroid belt. Yet it also comes to within 20 million km (13 million miles) of the Earth's orbit –

close in astronomical terms. Computer simulations show that Eros's orbit is stable over a long period measured in tens of millions of years, perhaps up to a hundred million years, whereas the lifetimes of most NEAs are more like several million years. But Eros is fated to crash into the Sun or a planet eventually, or otherwise disappear from our vicinity. Computer simulations by Patrick Michel and the late Paolo Farinella show that Eros has about a 5–10% chance of ending its existence by crashing into Earth a few million years from now. (It cannot strike us anytime soon.) If that happens, Eros would be the biggest object to hit since large lifeforms proliferated on Earth during the Cambrian epoch (about 500–600 million years ago). The resulting devastation would probably dwarf even the greatest mass extinctions recorded in the fossil record.

The scientific quest to learn about the nature and composition of Eros, not its potential role as a doomsday impactor, first motivated astrogeologist Eugene Shoemaker, of the US Geological Survey, to advocate a spacecraft mission to Eros. In the 1980s, Shoemaker hoped that Eros, as one of the largest NEAs, might contain extensive evidence about its genesis – perhaps a cross-section displaying layers that were once inside the larger, primordial parent body from which it was derived, like the rock layers visible in the sides of the Grand Canyon. Studying Eros's craters, topography, and structure was a principal objective in selecting NEAR's camera system. Other instruments were chosen to address the complementary issues of Eros's composition: whether or not it is made of materials like any meteorites we collect on Earth.

Before NEAR Shoemaker reached Eros, craters had been studied on images obtained at flybys of three other asteroids. Galileo imaging of the small asteroid Gaspra, in 1991, revealed an angular body, pinpricked with abundant small craters but few large ones. Two years later, Galileo's brief examination of a larger, main-belt asteroid, Ida, revealed surface features as small as 40 m. Ida looks different from Gaspra, much more like the Moon, with large craters at "saturation density". In other words, Ida's surface has roughly as many craters on it as will "fit". Subsequent impacts overlap and destroy preexisting craters as fast as they form new craters. Moreover, Ida looks like it may even be dominantly composed of two pieces, making it a kind of "rubble pile" asteroid, where the major fragments from a previous, not-quite-disruptive collision reaccumulated and remain loosely bound together by their mutual gravity.

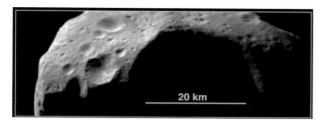

Figure 7.2. Looking into one of Mathilde's giant craters.

En route to Eros, NEAR Shoemaker encountered Mathilde, a large, charcoal-colored, main-belt aster-oid. NEAR Shoemaker's solar panels were designed to operate the spacecraft at Eros, rather than far out in the asteroid belt, so the power was sufficient to operate just the camera. Pictures of Mathilde reveal perhaps half a dozen enormous, gaping craters, each approaching or exceeding the body's radius – far more giant craters than have been seen on any other solar system body, including Ida and numerous planetary moons.

One might think, "seen one asteroid, seen them all," or at least that the cratered surfaces of asteroids should look the same. After all, the population of small, interplanetary projectiles is thought to have been roughly the same everywhere in the asteroid belt and in the inner solar system. So why should the crater populations on three asteroids look so different?

The NEAR Science Team pondered such issues as the spacecraft approached Eros to make much more comprehensive and detailed measurements than were possible during the brief flybys of Gaspra, Ida, and Mathilde. Mathilde's huge craters were thought somehow to reflect its very different composition. Its black color bespeaks its primitive, carbonaceous com-position, while the slight bending of NEAR Shoemaker's trajectory past Mathilde revealed that it has a very low bulk density – almost low enough for Mathilde to float on water! According to spectroscopic data obtained from ground-based telescopes, Gaspra and Ida have roughly similar compositions, which are also similar to that of Eros, yet they have different crater populations. Maybe Gaspra has fewer large craters because it was created by the break-up of its parent body comparatively recently and is younger than average, or maybe Gaspra is actually a difficult-to-crater metallic body, while Ida is more typically rocky. Would craters on Eros be like those of Gaspra or Ida, and why?

One difference between Eros and the other asteroids was *not* expected to lead to an observable difference in its geology. Since Eros's orbit is now wholly outside the asteroid belt, the present impact rate on its surface must be less than for main-belt asteroids like Ida and Gaspra by a factor of hundreds. However, even a hundred-million-year hiatus in impacts would be a tiny fraction of solar system history. Moreover, Eros's orbit might well have continued to stick out into the main asteroid belt during much of the time that it has been an NEA. So, either way, nearly all craters and other geologic features on Eros must reflect its accu-mulated history *in the main asteroid belt*, from the time it was created by a catastrophic break-up of its progenitor body, probably hundreds of millions to bil-lions of years ago, until it was perturbed on to its Earth-approaching orbit.

The larger craters on Eros

The premature flyby pictures showed not only the enormous saddle-shaped gouge, Himeros, but half-a-dozen other impact scars, including a 5-km -diameter bowl-shaped crater – since named Psyche – just oppo-site Himeros. From this early, distant perspective, it was hard to tell whether Eros was as saturated with craters as Ida, or more sparsely cratered, like Gaspra. In a way, Eros lacks the gigantic cavities of Mathilde; Himeros and Psyche are actually large compared with the *thickness* of Eros, but not compared with its length. One could imagine that, if an impact created Himeros, the cataclysm might have nearly broken Eros in two. Perhaps impact features can hardly be larger on such an elongated body.

Conceivably, Himeros may not be an impact crater at all but rather could be a kind of crease, formed when two chunks of what would become Eros bumped into each other. Or maybe it is a region damaged and excavated by shock waves that penetrated Eros from the opposite side when Psyche was created. It is intriguing that the enormous ridge Rahe Dorsum (named after the late Jürgen Rahe, who oversaw the beginnings of the NEAR Project from NASA Headquarters) rises from within Himeros and winds half-way around Eros to the vicinity of Psyche. Ridges and other linear features are very uncommon on the asteroids previously imaged by spacecraft. Yet Eros has a pervasive, global pattern of old, degraded, paral-lel ridges and grooves. Some researchers believe they may reflect a preexisting layered structure or "fabric" within the parent body from which Eros was derived. If true, then Eros may be a huge, shattered, but *intact* hunk of its parent body, rather than the jumbled rubble pile that would have resulted if separate pieces

Figure 7.3. A rogues' gallery of asteroids and probable once-asteroids
imaged by spacecraft missions and
compared at the same relative scale.
Upper left: the Martian moon Phobos;
Upper middle: 433 Eros; Upper right:
243 Ida; Middle left: the Martian
moon Deimos; Middle: 951 Gaspra;
Bottom: 253 Mathilde.

of its disrupted parent body had fallen back together again to make Eros.

Images of Eros taken shortly before NEAR Shoemaker's arrival on February 14, 2000, show a heavily cratered asteroid. Eros turns rapidly about its axis, about every 5.3 hours, bringing the sharply etched crater Psyche into view, then – half a rotation later – the great cavity, Himeros. The first couple of months of imaging from NEAR Shoemaker's initial, high orbit built up a portrait of Eros's sunlit northern hemisphere. (The southern hemisphere emerged into sunlight late in the mission.) Eros's crater population looks indistinguishable from that on Ida. In most places, craters larger than about 100 m across are pretty much cheek-by-jowl, reflecting a lengthy history of repeated impact-upon-impact, lasting perhaps two billion years.

Several places on Eros have fewer craters. Not unexpectedly, the steeply sloping interior walls of Himeros and Psyche are mostly devoid of craters; presumably, cavities that form on such steep slopes are quickly erased by cascading landslides, perhaps triggered by "Eros-quakes" due to later impacts. Another place with only one-tenth the density of large craters compared with more typical areas is dubbed Shoemaker Regio and lies southwest and equatorward of Himeros. Apparently, this misshapen region, littered with large boulders, was created by a large, recent impact. Shoemaker Regio's nature is still not thoroughly understood, but it is probably of rather recent origin, with inadequate time for subsequent impact craters to have re-saturated it.

Early, high-orbit views of Eros could not prepare us for what we would see during later months, when the spacecraft was lowered to an orbit just 35 kilometers from the center of the asteroid. For the moment, though, it was comforting that Eros looked like Ida and that both appeared rather like the familiar surface of the Moon. In the aftermath of Apollo, many studies were published about the lunar craters and "regolith" – that cratered, dusty "soil" on to which Neil Armstrong stepped on July 21, 1969. The Moon and its regolith thus became the archetype for the character of an airless world's surface, subject to cosmic bombardment.

Cosmic projectiles are not all the same size, like apples in a bushel. The numbers of large asteroids compared with smaller ones (and the smaller ones compared with tinier ones) can be described mathematically by a power law; the numbers increase rapidly with decreasing size, like the sizes of fragments of a brick smashed by a sledge hammer. Impactors make craters in proportion to their own size. So, for each kilometer-diameter crater (a little smaller than Meteor Crater in Arizona), there are hundreds of craters one-tenth as big (the size of a football field), and maybe a hundred thousand the size of a small house. The rain of ever more numerous, smaller meteoroids has literally sandblasted and pulverized the lunar surface repeatedly for billions of years. The resulting landscape is covered by craters of all sizes, right down to pits the size of the pebbles and grains that comprise the regolith. Indeed, microscopes reveal that the surfaces of returned lunar rocks are peppered with minute "microcraters", formed when invisibly tiny interplanetary dust particles zapped the Moon.

The mathematics of fractals reasonably well describes the lunar crater population: its appearance is similar whether you look at its biggest craters through binoculars, study astronauts' pictures, or examine returned samples through a microscope. As meteoroids and asteroids collide with each other, each object, acting as target, intercepts countless smaller objects. The fragmentation that results replenishes the supply of small particles, even as they are, in turn, being destroyed by still smaller ones.

Evidence of pervasive cratering of planetary surfaces is disturbed, of course, by active geological processes on planets like Earth, which erase craters, or by thick atmospheres, which prevent smaller craters from being formed in the first place (because meteoroids burn up in the atmosphere). But it seemed inevitable that close-up views of the surfaces of airless, geologically dead asteroids would resemble the pock-marked lunar surface. To be sure, asteroids differ

Figure 7.4. NEAR/MSI image of cratered and grooved terrain on Eros, with the Empire State Building in New York City shown for reference at the same scale. Several grooves (arrowed) are pointed out in this scene.

somewhat from the Moon: their weaker gravity ensures that some material ejected from craters will travel far around the body – some of it even escaping into interplanetary space rather than landing nearby, as on the Moon. Escape velocity from Eros is only 3 to 17 m/s so any kid could throw a ball off Eros into its own orbit around the Sun. As some ejecta are lost to space, the residual regolith is thinner than on the ejecta-retaining Moon, so impacts are more likely to excavate fresh rocks from below rather than repeatedly churn and re-pulverize the same, localized ground cover. Such differences between the Moon and asteroids were expected to appear subtle. We were wrong.

Boulders galore! (but where are the craters?)

It was in mid-March 2000, that the NEAR mission spacecraft was named NEAR Shoemaker, in honor of Gene Shoemaker, who had died shortly after attending NEAR's Mathilde flyby activities at APL. Soon afterwards, engineers moved it into progressively lower orbits at heights of 100 km, 50 km, then 35 km from the asteroid's center of mass. The closer images soon revealed Eros with a clarity never before achieved for any asteroid. With features as small as 10 m across resolved, Eros looked at first glance like the Moon. But there was a subtle difference! Most of the smallest features were actually bumps, not craters. With sunlight shining from the right, craters could be recognized by the fact that their interior walls were illuminated on the left and in shadow on the right. It was the *right* sides of most small features on Eros that were illuminated, and they cast shadows to the left. The larger ones clearly were enormous boulders, tens of meters across or more, but, at this point, it was too early to be sure that the smaller ones were true "boulders". Their shapes were demarcated by only a few picture elements, so we used the neutral term "positive relief feature," or PRF, to describe the mounds, hills, protuberances, blocks, or whatever they might be.

If most small features in the low-orbit pictures were PRFs, where were the small craters? Shockingly, with few exceptions, they are not there! Everywhere we look on Eros, craters 10–20 m across are rare; craters smaller than this are scarcer still. By analogy with the Moon, Eros's surface was supposed to be just as saturated with smaller craters as with large ones; instead, there is a dearth of craters below about 100 m in size and the shortage gets more marked for smaller craters. Craters several hundred meters across cover 25% of Eros's surface – as many as can realistically fit, given that larger and smaller craters must fit as well and that each new crater destroys some preexisting craters nearby. Fifteen-meter craters cover just 1% of the surface. Indeed, there are more boulders less than 15 m across than craters of similar sizes!

I presented these perplexing results at the annual meeting of the Division for Planetary Sciences of the American Astronomical Society in Pasadena, California, on October 23, 2000, just a couple of days before NEAR Shoemaker had an even closer look at Eros. The operations team at APL had grown more confident that they could control the spacecraft's orbital path in Eros's irregular gravity field so they allowed it to zoom to within only 6 km of the surface in a so-called low-altitude flyover (LAF). In my talk, I speculated about what the pictures might show later in the week. At the rate that smaller boulders were beginning to dominate the landscape on Eros, an extrapolation to still smaller sizes would have bus-sized boulders literally piled on top of one another everywhere. But in science, you cannot trust extrapolations: you must look.

I downloaded some LAF images over the weekend. Indeed, there were still more boulders, but car-sized blocks were not piled all over. They have varied

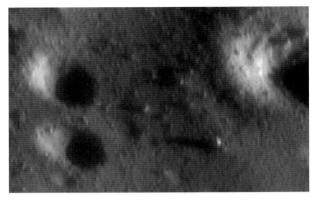

Figure 7.5. An example of an image from the first NEAR low-altitude flyover (LAF) maneuver, showing boulders and other positive relief features at very high resolution. The smallest resolvable features in this image are only about 1 m (3 feet) across.

shapes, meriting more precise descriptions than "PRFs". Most are enormous, angular rocks, the size of anything from an automobile to a house, some perched on top of the surface, others partly buried. Presumably, if we can go by what happened on the Moon, most arrived after being hurled from large, faraway impacts, such as the events that created Psyche and other large craters. Some may even have been in temporary orbit around Eros, finally angling into the surface again. Landing at the speed of a cyclist, such rocks and ejecta should bump and slide across the surface, leaving scars and trails in the regolith; blocks arriving on more vertical trajectories might plop into the surface, finally resting in a small cavity. Yet evidence of such touchdowns is scant. Very few boulders are found within dimples or at the ends of trails and there are alternative explanations. The blocks may be fractured components formerly in the interior of Eros and its parent body, which have gradually been exhumed from below as the asteroid's surface has been eroded over the aeons. Seen more clearly, a small proportion of PRFs look like huge piles of dirt or fractured rock but one cannot say whether they broke upon landing, or were shattered by a hypervelocity meteoroid after a re-impact or exhumation. A few PRFs, shaped like tilted slabs or rock towers, look more like outcrops of bedrock than ejecta tumbled in from afar, which should come to rest lying flat.

In order to understand Eros in a quantitative way, I opened software that I use to measure and tabulate craters and boulders. Written by Jonathan Joseph and Peter Thomas of Cornell University, the code resides on a computer in the NEAR Science Data Center at APL,

Figure 7.6. An example output screen from the software used to analyze craters and boulders in NEAR MSI images. Literally thousands of craters and boulders and other features were analyzed in terms of their locations, sizes, and shapes.

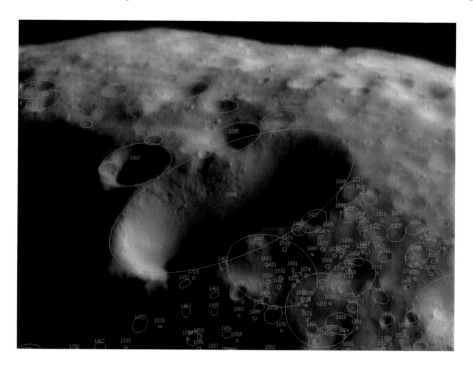

which I access via a secure protocol from my office in Boulder, Colorado. I select an image of interest, then use my mouse to mark the peripheries of a crater or boulder. The program has access to all picture parameters – when it was taken, the spacecraft's distance from the center of Eros, the science team's latest model for Eros's shape (hence the distance and tilt of the locality I am studying), the length on Eros's surface represented by a picture element, and so on. The software calculates the feature's diameter as well as its latitude and longitude. Despite the code's complexity, internet connections are fast enough to permit me to work at my natural speed from my office half way across the country from APL.

The PRFs were so numerous that it took me hours to measure hundreds of them from just a quarter of one picture. Yet when I measured the smaller craters, I was finished almost before I began: only a handful of small craters were in the entire image! The smallest-scale features, only a few picture elements across, are tricky to assess. The images have unavoidable artifacts; for example, like a cat's vision through its slit-shaped pupil, sharpness is unequal in the horizontal and vertical directions because of the unusual rectangular picture elements in NEAR Shoemaker's CCD camera. Moreover, the brain may interpret as a crater what is simply a small space between two irregularly shaped PRFs. One thing became clear: at least in the LAF images I studied, there were nearly a hundred PRFs for each crater of the same size! And this LAF strip turned

out to be in a comparatively boulder-free part of Eros! So, why are there so few small craters?

Logically, there are only three ways to explain the paucity of small craters. Most directly, there may be a scarcity of small projectiles to make small craters. Secondly, it may be that when small projectiles strike Eros's surface, no crater results (for example, if Eros were armored by a completely impervious surface). A final possibility is that small craters were formed, and continue to be formed, but some process (filling, erosion, collapse) has been very efficiently erasing them.

The last explanation is the most natural one. It is easy to imagine how to make a crater vanish: fill it in, cover it up. That is what is happening before our eyes to Arizona's Meteor Crater, as thunderstorms year after year, millennium after millennium gully its interior and wash debris on to its floor. It is what is happening (or happened) on Mars, as water and lava poured across its surface in the past and as winds blow sand and dust about today. Even on an airless, geologically dead world like the Moon, cratering itself erodes and covers up preexisting craters, especially when impacts have gone on long enough to reach saturation. But in the well-understood, lunar-like way, there should remain enough craters of every size for saturation coverage all the way down to the smallest.

This explanation simply does not work for Eros. There are only 1% of the number of 4-m craters on Eros expected under circumstances where cratering is

Figure 7.7. The spatial density of
craters and boulders on Eros as a
function of their diameter, ranging
from 1 cm (about 1 pixel on the final
descent image) to 10 km (roughly the
diameter of Himeros, the largest crater
on Eros). R is the fraction of area
covered. The scales on both axes are
logarithmic.

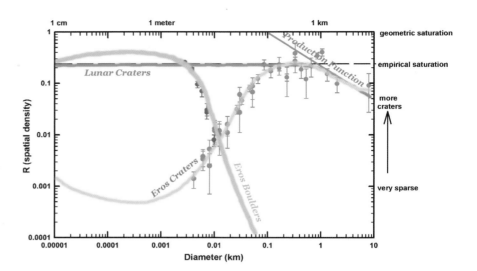

At the top of the graph (R = 1), the
area of craters near a particular size
would equal the area of the surface on
which they were counted; in realistic
cases of "saturated," heavily cratered
terrains on the Moon, planets, and
satellites, the maximum spatial
density has been found empirically to
be between 0.2 and 0.3 (the dashed
line). Craters larger than 1 km in size
are under-saturated on both Eros and
the lunar maria, but in the case of smaller sizes where the "produc-
tion function" of impactors would make more craters than can fit on
to the surface (red line), older craters are overlapped or destroyed
and we see only younger craters, which follow the empirical saturation density down to much smaller sizes, like craters on the lunar
maria, shown by the green curve.

However, on Eros, craters with diameters smaller than about
100 m are seriously and unexpectedly under-represented. At diameters less than several meters (with densities below the "very sparse"
level), craters are so rare (only a few on an entire NEAR Shoemaker
image) that it is difficult to estimate their spatial densities accurately. In contrast, there is a remarkable abundance of boulders on
Eros measuring under several tens of meters, and there are no boulders larger than 150 m. Going from larger to smaller diameters, their
density increases dramatically, then levels off at sizes smaller than a
couple of meters, before becoming so numerous as to be literally
piled on top of each other. Boulders are hundreds of times more
numerous than craters of similar sizes in the highest-resolution
NEAR Shoemaker images of Eros.

Representative counts, with error bars, are shown as red symbols
for craters and blue for boulders. The curves are schematic; in
reality, the spatial densities of both craters and boulders vary by
roughly a factor of 10 from place to place on Eros.

saturated. No filling or covering process can so dramatically deplete small craters while having minimal
effects on larger ones. Consider a fictitious example of
filling craters with dust or water, evenly deposited
from above. Craters of all sizes have roughly the same
proportions, so a crater ten times smaller than a larger
one is one-tenth as deep; it will take one-tenth as long
to fill up, so we should expect to see one-tenth as
many of them. On Eros, however, there are *hundreds*
of times fewer craters one-tenth the sizes of larger
ones, not just one-tenth. What process can single out
and destroy small craters so effectively, leaving larger
ones intact? There is none that I can think of.

Another stark piece of evidence further rules out
blanketing or erosion as an explanation for the rarity
of small craters. Boulders should be covered up or
eroded too! To be sure, boulders of a particular size are
typically several times taller than the depth of a crater
with the same diameter so it should take several times
longer to bury a boulder than fill a crater, or several
times longer to erode away their greater volume. But
that does not approach the disparity of a factor of one
hundred, nor the radically different size trends. We are
forced to seek another solution.

Consider the second option for explaining the
paucity of small craters: the meteoroids hit Eros, but
they do not make craters. Why could that happen?
Eros is not made out of diamond or some kind of
impervious kryptonite. From the perspective of an
incoming hypervelocity projectile, what is possibly
different about ground-zero on Eros compared with a
typical spot on the Moon? For one thing, all those
boulders! Instead of explosively excavating a small
crater in the lunar regolith, a projectile is more likely,
on Eros, to strike a big boulder. Its energy and momentum could fragment and destroy the boulder without
cratering the surface. Pervasive boulders might
somehow armor the surface of Eros. There are two
problems with this explanation. First, we must explain
not only the paucity of craters but also the abundance
of boulders, yet this process *destroys* boulders.
Conceivably, there is an inexhaustible re-supply of
boulders (for example, if boulders are being exhumed
from below as rapidly as they are being destroyed).
The second problem is more direct: despite their
unprecedented abundance , boulders do not literally
cover Eros. There are plenty of spaces in between for

projectiles to make craters. Indeed, the October 2000 LAF images revealed some perplexing places where there are neither boulders nor craters! (Even more striking examples were imaged in later LAFs, which were executed two weeks before the end of the NEAR mission.) They were called "ponds" – not literally water ponds, of course, but very flat, sharply bounded tracts in topographic lows, as if someone had poured in a fluid that had solidified. Whatever the ponds are, they contain few boulders and present an excellent, non-armored target for small cratering impacts. To be sure, a few tiny craters can be seen in some ponds, perhaps more than on some neighboring "beaches." Nevertheless, while Eros's numerous boulders may prevent some craters from being made, the principal explanation for the comprehensive depletion of craters must be something else.

We are left with one final option: there must be a deficit in small projectiles. But I will return to that.

Eros *very* close up

The NEAR spacecraft was never explicitly designed to land on Eros. It was not equipped with landing gear or feet. But the mission had to end somehow and Mission Director Bob Farquhar always dreamed of landing. He knew that the spacecraft was built to be tough: it had to survive the buffeting of launch, after all. And he knew that Eros had minimal gravity, so NEAR Shoemaker would hit at a speed of only 2 m/s, more gently than the Vikings landed on Mars. However, NASA officials were afraid to raise public expectations of a successful landing and averse to using emotionally laden words like "crash landing" to explain what was to happen. They tried to finesse such PR difficulties by mounting a press conference: the NEAR mission was lavishly declared to be a success two weeks before the landing. And, it was ordered, in any discussion of the activities to come after February 12, 2001, that the landing attempt would be described as a "controlled descent toward the surface."

The descent promised to be richly rewarding, scientifically. Since NEAR Shoemaker's original 200-km orbit, each closer, clearer view had divulged attributes of Eros never previously imagined. The final descent might yield a picture from 10 or 20 times closer than even the latest, best LAF. Just before touchdown, the ground would be too close to be in focus and there would be inadequate time to transmit a last image before the high-gain antenna was eclipsed from Earth.

As it turned out, NEAR Shoemaker's final picture,

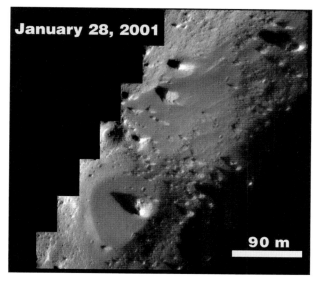

Figure 7.8. A mosaic of images of "ponded" materials seen in the January 28, 2001 low-altitude flyover (LAF) maneuver. These sharply bounded deposits appear remarkably smooth and are (at least to many) a wholly unexpected and new class of asteroid surface feature.

snapped from an altitude of just 120 m, is technically not in perfect focus but it is valuable, showing objects only a couple of centimeters across. The bottom third of the picture is, indeed, missing even though the spacecraft landed in a fully operational mode at 3:02 p.m. EST on February 12, 2001. Bob Farquhar would have liked NEAR Shoemaker to have taken off again, maybe to look down at any dents it had made in Eros's surface. Had it tipped upside down upon landing it *would* have taken off again on Valentine's Day, exactly one year after it first went into orbit: Farquhar had installed a command to fire the thrusters and do just that.

A "NEAR-print" might have revealed just how hard or soft the surface of Eros is. But Farquhar's wished-for maneuver would probably have failed. NEAR Shoemaker would have hopped up again, no doubt – a fine stunt. But there would not have been time or fuel to orient the spacecraft, find and image the first landing spot, and transmit a picture to Earth. And there were more important things to do. The gamma-ray instrument, which had barely even detected Eros during the year-long orbital mission, was now nestled in the dirt – an optimum location for the still-functioning instrument to belatedly accomplish its goals. So, the engineers, who remained in contact with the spacecraft, disabled the "take-off" command and NEAR Shoemaker remained in its final resting spot, taking data for two more weeks.

Figure 7.9. The fifth from last image acquired during the NEAR Shoemaker's final descent to the surface of Eros on February 12, 2001. The largest positive relief features in this rough, rocky-looking scene are about 3 m (10 feet) across.

Figure 7.10. The final image acquired by the NEAR MSI instrument, moments before the spacecraft touched down on the surface on February 12, 2001. The image is approximately 6 m across, and the resolution is approximately 1.2 cm/pixel. For scale, and to give a better sense of perspective to this amazingly high-resolution spacecraft image, a basketball (24-cm diameter) has also been fancifully (and gently) placed on the surface of the asteroid.

Meanwhile, we puzzled over the last pictures. Some show a rockier surface than we had ever seen from farther away. One could hardly leave a footprint clambering across such a rough landscape. But the very last pictures apparently cross into flat terrain, possibly one of the so-called ponds. The final image shows an area no larger than a small room. At the top is part of a desk-sized boulder. But most of the locality is eerily flat, without many rocks or pebbles and quite lacking in craters. There is one little, curvaceous "canyon," such as a child might scrape into beach sand, with rims maybe a couple of centimeters (an inch) high. It is a mystery. At closest range, Eros looks more like Mars than our prejudices of a meteoroid-pitted asteroid.

A conundrum solved?

With hindsight, the answer to the many-boulders/no-craters conundrum may be stunningly obvious in its simplicity. The same week that the October LAF pictures were received, an issue of the international journal *Nature* was published containing an essay I had written describing a new paradigm to explain how meteorites reach Earth from the asteroid belt, which could prove to be the answer. Unfortunately, I had compartmentalized my essay in one part of my brain while another was struggling to understand the close-up appearance of Eros. The connection was made a couple of months later by Jeffrey Bell, an asteroid researcher at the University of Hawaii.

Back in the early 1990s, while NASA was beginning to implement FBC and get the Discovery program underway, a pivotal conference was held in San Juan

Capistrano. NASA chief Dan Goldin attended and gave an inspirational talk to would-be proposers of future Discovery missions, touting those already in the pipeline: the Pathfinder Mars mission and NEAR. During the question-and-answer session, Jeffrey Bell rose to challenge Goldin about NEAR. NEAR was too cheap and inadequately equipped to *really* solve the mysteries of Eros, Bell contended. NEAR was a waste of money, he told Goldin.

But in January 2001, as the NEAR mission came to its spectacular end, Jeffrey Bell could not help but marvel at the close-up images posted on the NEAR web site. Why were there so many boulders and so few craters, he wondered? Suddenly, it came to him: there could be a paucity of projectiles in the size range capable of making smaller craters or destroying the boulders! An unconventional idea like that might well occur to an iconoclast like Bell. It violated the long-held paradigm, dating back at least to a 1953 paper by S. I. Piotrowski. Since that time, generations of researchers have accepted the idea that, after aeons of mutual catastrophic collisions, interplanetary projectiles should be increasingly abundant with decreasing size in accordance with a power law. They assumed the rule applied both within the asteroid belt and to the dribble of meteoroids that escapes to Earth and the Moon.

The chief exception to the power-law rule was at very tiny sizes, where solar radiation effects, like light pressure, effectively blow away, or drag in, dust-sized particles. More recently, however, as I wrote in my *Nature* essay, new work by David Vokrouhlichy (from Prague) and by Paolo Farinella[1] (from Pisa) proved another exception for much larger objects. Their paper capped sporadic development of an idea first presented in a now-lost paper, published around 1900 by an East European engineer, I. O. Yarkovsky. As Vokrouhlichy and Farinella's paper (and my essay) explain, solar heating of a spinning meteoroid, and cooling by re-radiation, cause its orbit to drift because "afternoon" or "summer" warming does not exactly cancel out "morning" or "winter" cooling. Over millions of years, hand-sized and boulder-sized bodies drift with respect to the more stable orbits of larger asteroids, until they enter a so-called "resonant escape hatch" and are chaotically extracted from the main asteroid belt into elliptical, planet-crossing orbits. The last phase of this process is similar to the way Eros itself escaped from the asteroid belt but Yarkovsky forces barely affect bodies as large as Eros. In the case of Eros, slower-acting dynamical forces, and perhaps even the kick it received when its parent body broke up, moved Eros into the "escape hatch."

Specialists in interplanetary dynamical processes and in meteorite cosmic-ray ages accept that Vokrouhlichy and Farinella's computer modelling of the Yarkovsky effect, finally and satisfyingly resolves the long-standing problem of how we get meteorites to the Earth from the asteroid belt. For decades, there had been hand-waving suggestions, but closer looks revealed quantitative difficulties. Now, we realize, there is an efficient, inevitable mechanism that removes fragments sized between centimeters and tens of meters from everywhere in the belt and delivers some of them to Earth as meteorites.

Jeffrey Bell recognized the consequence for the size distribution of bodies in the asteroid belt: that *there must be a deficit of meteorite-sized asteroids.* As it turns out, such objects have just the sizes that would make the craters that are "missing" on Eros! And, despite Eros's current location outside the asteroid belt, dynamicists believe that the vast majority of its surface features (including boulders and small craters, or lack thereof) must have been formed during the first

99% or so of its history, while it was still in the asteroid belt. With hindsight, there is no wonder that Eros and the Moon look so different. The Yarkovsky effect removes small bodies from the asteroid belt but enhances the number of them available to strike Earth and the Moon.

As Jeffrey Bell and I followed the developing literature about the Yarkovsky effect during the 1990s, we had the knowledge to predict that Eros would have few small craters and many boulders. Yet it often takes real new data to stimulate the theory and the intellectual connections necessary to understand our universe. Jeffrey Bell may or may not have made a valid critique of NEAR's deficiencies when he lectured Dan Goldin. The irony is that he, and we, needed NEAR – imperfect or not – to venture forth and see what Eros is truly like. We dare not rely on theoretical speculations in the absence of data. Asteroids are small and faraway. So we must go out and explore them. NEAR Shoemaker was the harbinger of many voyages to come.

Figure 7.11. Just like on the Moon and many other solar system bodies, craters tell a major part of the story of the history of Eros. This mosaic of four images taken by NEAR Shoemaker on September 21, 2000, from about 100 km (62 miles) above Eros covers part of the asteroid's southern hemisphere, southwest of Psyche. The ridge that trends from upper left to lower right is probably among the older features on Eros, as evidenced by the large number of superimposed impact craters. The whole scene is approximately 11 km (7 miles) from top to bottom.

---

[1] Farinella was perfecting details of his own paper when he was rushed to hospital for a heart transplant just as NEAR Shoemaker began its year in orbit around Eros. The NEAR Science Team mailed a large autographed picture of Eros to Farinella, but he never regained consciousness from the operation.

8 On Course and Picture Perfect

Maureen Bell
Cornell University, Ithaca New York (NEAR Sequencing Team member)

Bill Owen
Jet Propulsion Laboratory, California Institute of Technology (NEAR Navigation Team member)

As one of NASA's first Discovery missions, NEAR was designed to be "better, faster, cheaper", hopefully to fulfill the program's motto. It took five years from launch to completion of the mission, which ultimately was successful on all counts. It thoroughly investigated an asteroid millions of miles away with many different scientific instruments, controlled by small groups of people distributed across the United States. Its success depended on the ultimate in teamwork: it takes more than rocket science to make a spacecraft do what you want it to do.

Navigation: How we get there from here

Primary responsibility for the navigation of any spacecraft lies with the mission's navigation team, or "NAV" for short. NEAR's NAV team was based at the Jet Propulsion Laboratory (JPL) in Pasadena. They did the math of figuring out where we were and where we wanted to go. Periodically, NAV created a computer file, called a trajectory file or ".bsp file," that contained this information and which provided all of the impor-

tant spacecraft navigation data to the other teams. But NAV also relied on other teams within the mission to communicate with the spacecraft.

The Mission Operations team (MOPS) and Mission Design team (MD) at Johns Hopkins University Applied Physics Laboratory (APL) had the task of taking the trajectory files with those all-important numbers and transmitting them up to the spacecraft so NEAR Shoemaker would know where it was and what to do. The tasks included designing rocket burns, maintaining the health of the spacecraft (checking whether the solar panels still pointed toward the Sun, for example) and programming the daily return of scientific data and other information from the spacecraft.

The multispectral imager team (MSI) at Cornell University had the task of designing special commands that would use NEAR Shoemaker's camera to collect images and aid the NAV team in determining the spacecraft's position. The images, known as "opnavs" (short for "optical navigation images"), were collected periodically throughout the mission. Opnavs

Figure 8.1. The Cornell University NEAR Shoemaker spacecraft sequencing team. Clockwise from lower left: Maureen Bell, Elaina McCartney, Jonathan Joseph, Colin Peterson, Brian Carcich, and Ann Harch.

Figure 8.2. The Jet Propulsion Laboratory/Caltech NEAR spacecraft navigation team, celebrating the spacecraft's successful arrival in Eros orbit. Front row, left to right: Steve Chesley, Tseng-Chan "Mike" Wang, Jon Giorgini, John Bordi. Back row, left to right: Jim Miller, Bobby Williams, Pete Antreasian, Cliff Helfrich, Bill Owen, and Eric Carranza.

Figure 8.3. The Johns Hopkins University/Applied Physics Laboratory NEAR mission operations team, celebrating the successful completion of the mission. Front row, left to right: Lisa Segal, Carolyn Chura, Pat Hamilton. Middle row, left to right: Ron Owen Dudley, T. J. Mulich, Mark Holdridge (team leader), Nick Pinkine, Rolland Rolls, Dina Tady. Back row, left to right: Bob Dickey, Karl Whittenburg, Rick Shelton, Bob Nelson, Jon Rubinfeld, Charles Kowal, and Charles Hall.

of stars were obtained and analyzed during the Mathilde flyby and the cruise to Eros, and then opnavs of Eros itself were obtained and analyzed during the approach and orbital phases of the mission.

At its most basic level, navigating a spacecraft around the solar system is mostly about answering two questions:

1   What is the current course of the spacecraft? We need to know not just where it is at a particular moment of time, but also how fast it is going and in what direction, in order to be able to predict its future trajectory.

2   If the answer to question 1 is significantly different from the desired flight path, then we need to ask: How do we get the spacecraft back on to its correct trajectory? We need to command the spacecraft to fire its thrusters and perform "trajectory correction maneuvers," to nudge it back toward where we want it to go.

The two main tasks of spacecraft navigators like those on the NEAR NAV team are to mathematically determine the spacecraft's orbit, and to design any needed course correction maneuvers. These tasks require the application of celestial mechanics, numerical analysis, filter theory, and – yes – rocket science. We will describe first how the process usually works and then the changes from the usual pattern that had to be made for NEAR.

Orbit determination, or OD for short, is the art of using measurements such as radio tracking, pictures, or altimeter ranges, to improve upon our estimate of a spacecraft's trajectory. We have to accept that we can never claim to know the trajectory of a spacecraft exactly. Our observations are not perfect, and a

multitude of similar trajectories can fit them adequately. Each time the NAV team generates the official "trajectory", it is actually just the one most likely to be correct, but there is always an element of uncertainty in both position and velocity. If we think of the optimal, desired trajectory as a thin curve between two points in space, then the best we can do to determine the actual trajectory of the spacecraft is to imagine it within something like a fuzzy tube that surrounds that curve. Furthermore, the size and shape of the fuzzy tube can vary. The tube is small in places where we have good data to constrain it, but it gets fatter as one tries to predict the trajectory farther and farther into the future.

The observations used by NEAR NAV for the normal orbit determination were radio ranging and Doppler from ground-based tracking antennas and optical navigation pictures from the on-board MSI camera. The acquisition of range and Doppler data relies on the tracking antennas of NASA's Deep Space Network (see box on following page). Optical observations are made with the spacecraft cameras.

In order to obtain range data, a DSN antenna must transmit a special set of modulated signals to the spacecraft, in essence similar to an FM radio signal. The spacecraft's on-board computer, which receives the transmission and recognizes it as a special signal from home, has software to retransmit it back to Earth immediately. Electronics at the DSN station determine the time when the return signal reaches the antenna back on Earth, and compare it to the known time that the signal was transmitted. The time difference is the time that it took the signal to get to the spacecraft and back at the speed of light, or the "round-trip light

The Deep Space Network

NASA's Deep Space Network (DSN) consists of a series of radio telescopes (antennas) near Goldstone in California, in Spain near Madrid, and near Canberra in Australia. Each DSN site has one 70-m radio dish and several smaller antennas at each location. The three sites are distributed around the Earth fairly evenly in longitude so that it is possible to receive information from a spacecraft (downlink) and send information to a spacecraft (uplink) wherever they are in the solar system without interruption as the Earth rotates. The DSN is the primary way that NASA sends and receives signals from the dozens of interplanetary space probes traveling throughout the solar system and beyond.

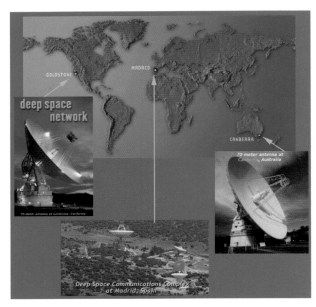

Figure 8.4. The Deep Space Network consists of three 70-m radio telescopes and a number of 34-m radio telescopes spaced approximately evenly around the world allowing constant communications with deep-space missions.

time," and is a direct measure of the distance from the DSN antenna to the spacecraft. The one unfortunate characteristic of range data is that the spacecraft cannot transmit science data very efficiently when it is sending back the range code. We therefore used range data rather sparingly, typically for just a few brief ranging sessions per week.

Doppler data are a measurement of the shift in the frequency of the received radio signal relative to the known frequency of the originally transmitted signal. Anyone who has ever watched a train pass by has experienced the Doppler shift of sound: the train's whistle is high-pitched as it approaches you, and then

changes to lower-pitched after it goes by. The whistle itself is constantly emitting the same tone—our perception of the tone changes because the train is moving relative to us. The same kind of Doppler shift occurs with light: radio waves (which are just a form of light) from a spacecraft moving towards us are shifted to higher frequency, and those from a receding spacecraft are shifted to lower frequencies. For NEAR Shoemaker, we preferred to use a technique called "two-way Doppler" in which a signal is transmitted from the DSN at a known frequency, received and retransmitted by the spacecraft in the same way as the range data, and finally detected back on the ground. The DSN measures the incoming frequency very accurately – to a fraction of a cycle on a signal with over 7.6 billion cycles every second. Because the antenna and the spacecraft are in relative motion, there is a difference between the final received frequency and the original frequency – the Doppler shift. The size of the Doppler shift depends on the relative velocity between the spacecraft and the antenna in the direction of the line joining them. The antenna happens to be sitting on a rotating Earth, so the first part of the Doppler shift arises from the antenna's own motion. This is a valuable effect to exploit for additional information, because it depends on exactly where in the sky the spacecraft happens to be. Another part of the Doppler shift arises from the Earth's orbital motion around the Sun; this is not as interesting from a NAV standpoint, but we can compute what it is quite well. A third component comes from the spacecraft's own motion, and this is where things get interesting, especially for NEAR Shoemaker while it was in orbit around Eros. Whenever Eros's rotation brought one of the asteroid's long ends closer to NEAR Shoemaker, the spacecraft would feel a little extra gravitational tug. Similarly, when NEAR Shoemaker was over Eros's "waist," it would feel a little bit less gravity. This variable force produced a noticeable change in the spacecraft's orbit, wiggles that show up distinctly in the Doppler data. This is how we determined the various numbers that describe Eros's gravity field, as well as the size, shape, period, and orientation

of NEAR Shoemaker's orbit. Doppler truly is the workhorse of the interplanetary navigator.

Obtaining and analyzing optical data to improve the trajectory are an entirely different matter. Here we use a camera on the spacecraft (the MSI) to take special pictures for navigation purposes. In all previous JPL missions, the NAV teams did their own sequencing, or at the very least gave to the sequence team a list of times and directions in which to point the camera. For NEAR it was different. It was the first time that a JPL NAV team delegated the picture planning to somebody else. The NAV team was in control of requesting all imaging time for opnavs, approval of any changes to them, and verification before each uplink that what MSI was being commanded to do was what NAV wanted. The system worked exceptionally well, for two reasons in particular: MSI had superior planning software (a program called *orbit* written by Brian Carcich at Cornell), and NAV simply did not have the time and manpower to spend on sequencing, given that they were analyzing hundreds of images every day.

For earlier spacecraft like Galileo and the Voyagers, opnav pictures would contain a satellite against a background of reference stars. We would identify the stars in each picture and, since their coordinates were well known, we would know exactly where the camera was pointing. Then the location of the satellite image in the picture told us the direction from the spacecraft to the satellite. String enough of these pictures

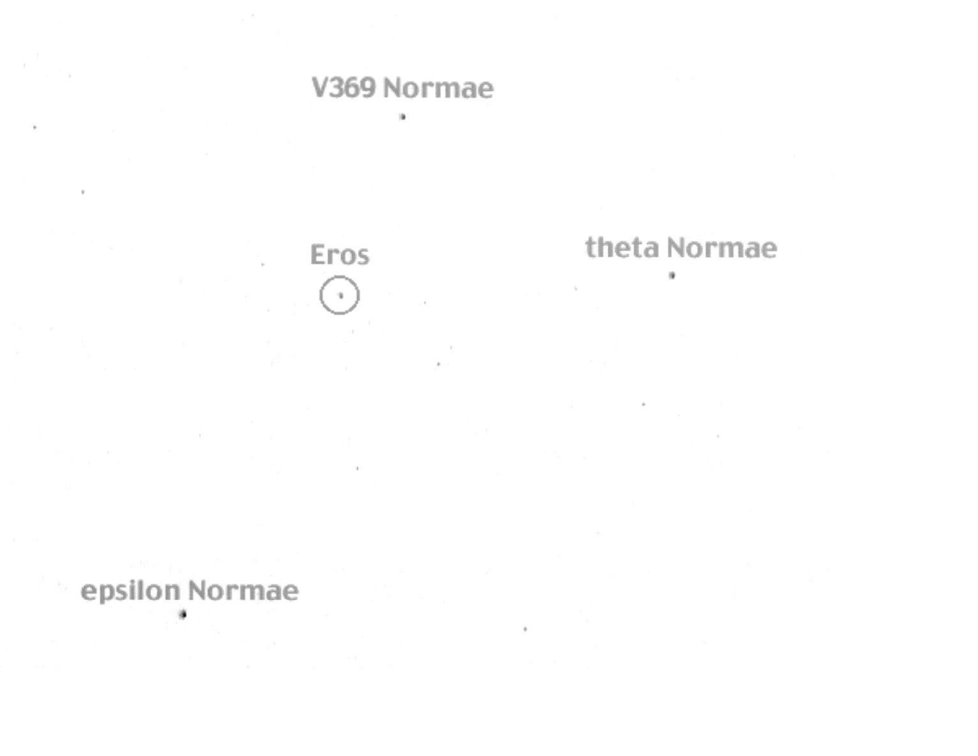

Figure 8.5. This first image of the asteroid Eros was acquired by the MSI camera on November 5, 1998, from a distance of 4 million km (2.5 million miles). Located at the center of this inverted image and circled, Eros appears against the star background as a single illuminated pixel. The exact location of the stars in the image helped to refine the trajectory so that the path to orbit insertion was exact.

Figure 8.6. Example of a NEAR MSI
optical navigation image of Eros on
which craters and other landmarks
were labeled and outlined in yellow
based on the predicted spacecraft
pointing (top), which was then refined
by aligning the landmarks to their
correct actual positions in the images
(bottom).

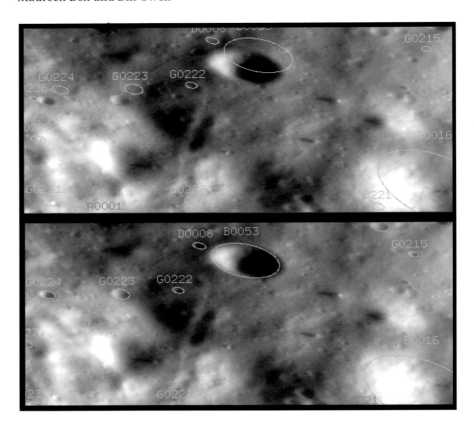

together and one sees the movement of the satellite
with respect to the background stars over time. In this
way one can determine both the spacecraft trajectory
and the satellite's orbit more accurately.

But the NAV team knew that traditional opnav
techniques would not work at Eros. The asteroid is so
irregular that its center cannot be located easily just by
looking at a picture. What is worse, its surface is quite
bright, so the camera is limited to short exposures, and
only the very brightest stars have any chance of
showing up. Therefore we had to develop an entirely
new approach that did not rely on stars or on finding
the center of Eros. Craters became the navigation landmarks for Eros instead. Not just any crater would do,
though. They had to be small craters with deep sides
and nice round rims so that they could be seen clearly
in all sorts of lighting conditions. They also had to be
easily identifiable; one misidentified crater could skew
the results. Fortunately, Eros has many thousands of
craters to choose from! There are over 1600 of them in
the Eros navigation database, and about a hundred of
the "landmarks" were used in operations over the
course of the year that NEAR Shoemaker orbited Eros.
Approximately 44 landmarks were used at a given
time during an orbit. As the mission went on and the
Sun moved progressively further south in Eros's sky,

the northern side of Eros became shrouded in darkness
and the southern regions, which had been in darkness,
became sunlit. Our choice of landmarks had to follow
the Sun. Stars were still an important tool, but only
indirectly; NEAR Shoemaker's star tracker (a separate
camera system dedicated to identifying the background stars) told us the orientation of the whole
spacecraft, and that in turn gave us the direction in
which the camera was pointing.

Measuring the landmarks was one of the few things
that was literally done by hand. Each picture of the
surface of Eros was displayed on a computer monitor.
The computer would use the expected trajectory to
calculate where the craters should be and put on top of
the picture a simple overlay drawing of the outlines of
the expected landmarks. First the cursor would be used
to drag the crater overlays until they lined up with the
actual picture. Then, using the cursor again, points
were selected all around the rim of each crater. The
computer would examine those points and find the
center of the crater. The difference between where the
crater was and where the trajectory thought it should
be told NAV the deviation of NEAR Shoemaker's real
path from the required trajectory. The NAV team
would go through this process each day, looking at
perhaps 50 pictures and measuring a dozen or so

craters in each picture. During the year-long orbital phase of the mission NAV obtained, and team members Bill Owen and Mike Wang analyzed, some 134 267 crater measurements in 33 968 pictures (out of a total of about 181 393 pictures taken). It was a tedious job at times, but well worth the effort.

The orbit determination process took all these observations – not just the crater measurements, but the Doppler and range information too – and figured out which trajectory (out of the many possible trajectories) was the best match to them. JPL NAV team members Jim Miller, Pete Antreasian, and Steve Chesley were the masters of this black art.

The process of using all the available information to figure out the most probable trajectory depended in various ways on many things. These included the position and velocity of the spacecraft relative to the center of Eros; the size and direction of any needed orbit maneuvers and other thruster firings; the strength and shape of Eros's gravity field; Eros's rotation rate and the direction of its north pole, which together provide the asteroid's orientation in space; and a host of other "non-gravitational accelerations" on the spacecraft, for example from solar radiation pressure and thermal imbalances.

Each of these effects was modeled in computer software in terms of one or more "parameters," quantities that are initially unknown but whose values can be deduced from the observations. Other parameters, for instance the locations of the landmarks, did not alter the trajectory itself but did affect our observations. These were included in the OD process as well. There were several hundred parameters in total, hundreds to thousands of optical measurements, and many thousand Doppler and range data points available. In order to find the best solution, our software first answered two questions:

1   How does each observation compare with what it would have been if our knowledge of the trajectory were perfect? These differences are known as "residuals" and the object of the OD solution was to minimize them.

2   How would each residual change if each of the solution parameters were changed? This information helped us get a handle on how robust each possible solution really was, compared to the many others that were also possible.

With the answers in hand, the OD program was able to figure out the combination of parameters that did the best job of making all the residuals small. The residuals never vanished, because no measurement is perfect. Each measurement was slightly in error, and associated with each measurement was a best guess as to how big that error was likely to be. These measurement uncertainties meant that the final OD solution was likewise uncertain in some degree (hence our fuzzy tube analogy).

The OD team's best estimate of the trajectory then went to the maneuver analysts, Cliff Helfrich and Mike Wang of the JPL NAV team. Their job was to design the "orbit correction maneuvers" or thruster firings that brought NEAR Shoemaker back to its desired trajectory. This was yet another problem for the number crunchers. If we fire the thrusters in a particular direction, where does the spacecraft actually end up? Space is three-dimensional, so we asked this question for each dimension. Comparing the three answers to the OD team's estimate told us what each component of the maneuver would do. Comparing the estimate to the nominal (that is, where the spacecraft ought to be) told us what must be done. Finally, then, we had enough information to work out the maneuver.

Doing one maneuver is easy, but we often had to design four or five at a time in order to get NEAR Shoemaker where it had to be in a month or two down the road. The approach outlined above tends to be unstable, since very small maneuvers can sometimes lead to large changes in the trajectory, and the problem ceases to be "linear" – if you double the maneuver, you will not necessarily get twice the result. The maneuver design team had to proceed almost by trial and error at times, moving in very small steps until they had converged upon an appropriate answer.

NEAR Shoemaker had several problems during its long journey, most of them technical and related to on-board spacecraft systems, instruments, or software. Through it all, the new and time-tested navigation techniques described above never failed. NEAR Shoemaker stayed on course at all times during its complex dance around the solar system because of the hard work and talents of the NAV team and their colleagues on MOPS, MD, and MSI.

The flyby

In December 1998 as we approached the asteroid for orbit insertion, we were all nervous. We were working with new software, firing engines we had not used in a year and performing a maneuver that had never been attempted – entering orbit around a small body with unknown gravity factors. Several tests were run on the sequence in the hopes of minimizing our chances for disaster. The science teams, in particular the imaging team, were asked to put together "contingency" sequences in case something went wrong. David Dunham, of the mission design team at APL, is said to have run all possible contingency sequences for the operations team so that they would know what to do if there were an abort at any part of the orbit insertion sequence.

We felt that we were prepared, but we were apprehensive. The day of the insertion burn we all waited for news. And then the unthinkable happened. During the main engine firing the spacecraft stopped communicating to the DSN. We had lost NEAR Shoemaker and no one knew why. During an incredibly tense 24-hour period, the operations team worked ceaselessly trying to make contact with the spacecraft in order to regain control and find out what went wrong. We had mostly given up hope when, much to our surprise, we received a call from Karl Whittenburg from the APL MOPS team telling us that they had regained contact with the spacecraft and that we had 12 hours to put together a sequence for a flyby: the spacecraft was not going into orbit but was instead going to fly right past Eros just before Christmas.

The recovery of the spacecraft was both amazing and incredibly lucky. Because of a premature shutdown of the main engine, NEAR Shoemaker had been put into a spin, spewing thruster gas as it tumbled through space. The on-board computer had a special set of commands to follow in the event that it did not hear from Earth for a particular period of time. But because the spacecraft was tumbling and its antennas could not "lock" on the Earth, those commands were not working. NEAR Shoemaker was trying to take a star tracker image, determine where it was, and burn a thruster to put it in a configuration where it could transmit to Earth. But because the spacecraft was spinning rapidly, that process took too much time, so it would burn the wrong engine and Earth would not be in the right orientation. Finally, by chance, the recovery routine worked. The spacecraft figured out where it was, pointed toward Earth, and begged for help. Karl Whittenburg, who had not slept much in the several days of frantic activity, sent the commands to control the spacecraft and, using the contingencies that David Dunham had created, prepared it for resuming its journey. The mission was almost lost before it really began, but the hard work and earlier contingency planning of the team involved saved the day. It is also humbling to realize that plain old luck probably

played an important role in averting this near-catastrophe. It is rare to get a second chance in the space exploration business.

Now the JPL NAV team had the most difficult task. They had no data to work with after the first split-second of the aborted engine firing. Once the MOPS team recovered NEAR Shoemaker, NAV still had no way of knowing exactly what had happened during the time that NEAR Shoemaker was out of contact. NAV basically had to start the process from scratch with only a day or two of tracking data to pin down the trajectory, and only a few hours to find the best solution. The range and Doppler data were the true heroes in this phase of the mission, giving NAV a decent (though not very accurate) trajectory to use for the sequencing of the flyby.

MSI decided that they needed to take mosaics of images covering everywhere the asteroid might be, to compensate for the larger-than-usual uncertainties in the trajectory. For example, if the navigation team said that the asteroid would be at a particular position plus or minus 1000 km, MSI would create a mosaic of images arranged in a square that would cover this "error ellipse." These mosaics would be repeated as quickly as possible, taking into account that the spacecraft was flying past the asteroid, the asteroid was rotating, and MSI would have to be slewing the spacecraft to get the mosaic. But MSI needed to make sure that they did not slew the spacecraft too quickly and smear the pictures. It took hours for MSI team members Scott Murchie from APL and Ann Harch and Maureen Bell from Cornell to put together all of the commands. Sometime around 8.00 p.m. on December 22 the sequence of camera commands was delivered to MOPS. They worked on it all night to prepare it for the spacecraft and sent it up, finishing just a few minutes before the sequence needed to begin executing. It worked perfectly, and the mosaics covered the asteroid almost exactly as planned. And, just as importantly, the successful firing of the main engine shortly afterwards placed us on track again for an orbit insertion with the asteroid, though one year later than expected. The images and other information about Eros were a wonderful Christmas present to the world, and provided us with crucial information about the asteroid that was used to plan the best possible orbital mission.

Cruising to the asteroid

We had one year to figure out whether we could make our lives easier the second time around. It became

Figure 8.7. Just prior to orbit insertion on February 13 and 14, 2000, the NEAR Shoemaker spacecraft executed a series of observations using the NIS instrument with the Sun high in the Eros sky – optimal for collecting spectroscopic measurements. The predicted locations of NIS "footprints" from one such sequence of Eros measurements (left) resulted in a global spectral map; one example spectrum from the NIS data (right) shows inflections and weak absorption features diagnostic of the minerals olivine and pyroxene on the surface of the asteroid.

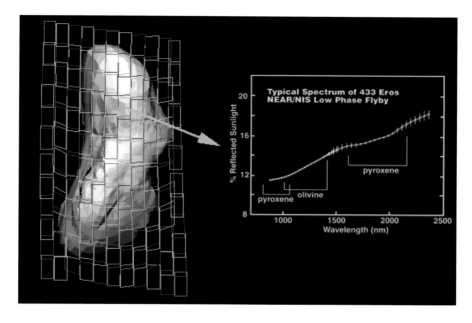

apparent during the flyby that MSI needed to have more sophisticated software – software that could not only compute the asteroid and spacecraft positions but also simulate the instrument commands. MOPS needed to work on a number of software problems on the ground as well as on the spacecraft. Everything needed to be fixed before January 2000 (not to mention any potential Y2K bugs) but it was decided not to send too many tests to the spacecraft during the cruise phase in order to conserve the limited available fuel.

The most important work for the imaging team involved the program that MSI used to do all of our sequence creation, *orbit*. The programmer at Cornell who had created *orbit*, Brian Carcich, spent many months adding the specific commands to move the spacecraft to create mosaics and the commands for the camera and NEAR Shoemaker's near-infrared spectrometer (NIS). One of the requirements was to be able to design something in *orbit* and create a file with the commands in the proper format so that they could be sent directly to MOPS – no edits or changes necessary. The MOPS team had all of the science teams create test sequences during the year and by December 1999, *orbit* was ready to carry the imaging sequencers through a year of weekly sequencing activities.

As we approached the asteroid again in January 2000, our software (both on the spacecraft and on the ground) was in the best possible state. We had a good understanding of each other as a team. We even had our approach sequences laid out. This time around we were not as nervous because the approach speed to Eros was only about 10 m/s, and slowing down to

begin orbiting Eros did not require firing an engine, but merely performing a series of small thruster burns like the ones that we had done several times over the cruise year.

One of the unique aspects of the orbit insertion was that we were turning the spacecraft so that we could observe the "low phase point" for the NIS team. The low phase point is the spot from which sunlight reflects directly back to the spacecraft. On water this would show the glint or mirror reflection of the Sun. On small bodies like Eros, obtaining data in this kind of configuration can provide unique information about surface properties, like particle size, slopes, and degree of compaction. It was a difficult maneuver but critical for achieving the NIS science goals; NAV had determined that the best opportunity to perform this low-phase flyby was during the first part of the orbital mission. NIS sequencer Colin Peterson from Cornell worked closely with NAV and the science teams to build and deliver an excellent sequence of low-phase spectroscopic measurements.

Orbit insertion and the high orbits

NEAR Shoemaker performed the low-phase flyby and the insertion maneuver without incident. We were happy that everything went well and were ready to begin a year of science observations – and what turned out to be navigational firsts.

The high orbits (200 and 100 km from the center of mass of the asteroid) were the highest priority for the imaging team. MSI needed to map the entire asteroid

Figure 8.8. The predicted footprints and Eros orientations for two representative MSI image mosaic sequences taken during the high-orbit (> 100 km altitude) phase of the mission (left), compared to the actual mosaics acquired from these sequences (right). These were the first two mosaics of Eros taken just after the engines fired and put the spacecraft into orbit.

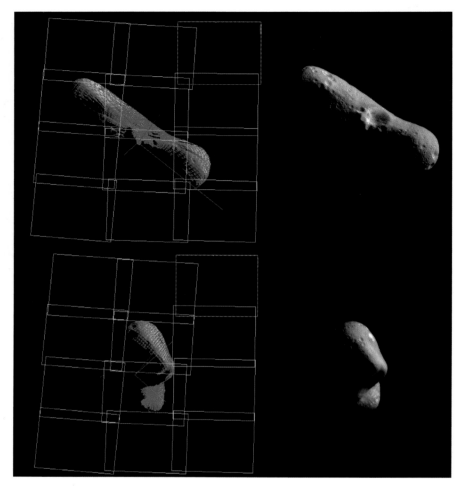

thoroughly from these distances through all of the available filters. We had not anticipated the amount of work that the weekly deadlines would entail and needed to continually readjust our workloads among team members.

The thruster maneuvers that changed the spacecraft distance for the high orbits were designed by Cliff Helfrich and Dan Scheeres on the JPL NAV team. They had calculated which orbits would be stable, in the sense that (a) they would not lead either to crashing on the surface or to escaping Eros's gravity and (b) small errors in the actual trajectory would not get magnified and lead to crashing or escaping. The maneuvers were usually only 1–2% less than or greater than expected. During these first few months it took the NAV team about a week's worth of data to come up with an accurate solution for the trajectory. Later in the mission, as the process was fine tuned, they could turn around a trajectory in a few days.

Additionally, MOPS had not anticipated some of the operating difficulties, including the need for extra trajectory uploads to the spacecraft for the many planned

observations. It was a period of adjustment for all of the teams and stressful for many of its members. Neither the spacecraft nor the asteroid cared about human concepts like weekends and holidays.

During these high orbits the opnavs provided a framework for the navigation of the spacecraft and a sanity check for the Doppler data, which provided the bulk of the navigation information. Optical data cannot produce the gravity field parameters as effectively as Doppler data, nor can they provide the absolute size of NEAR Shoemaker's orbit; if Eros were bigger but less dense and NEAR Shoemaker had been in a larger orbit for the same period, then the pictures would not look any different. However, without optical data as a third component of the analysis, it would have been extremely difficult to get the OD to converge on the right solution using just range and Doppler data. So opnavs were typically performed three or four times a day. The opnavs would attempt to cover the entire asteroid in mosaics. Some of the most spectacular imaging turned out to be these series of opnavs during the high orbits.

Figure 8.9. The predicted footprints and Eros orientations for two representative MSI image mosaic sequences taken during the low-orbit (50 km altitude) phase of the mission (left), compared to the actual mosaics acquired from these sequences (right). These pairs of image mosaics were taken at opposite ends of the asteroid within a 15-minute period three times per day.

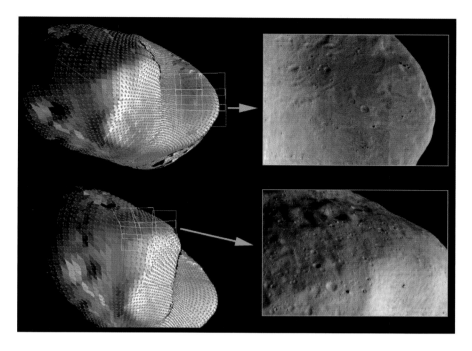

Low orbits

During the low orbits the imaging team was no longer in control of the spacecraft. The X-ray/gamma-ray spectrometer (XGRS) team decided what part of the asteroid we would observe during each day. The 50-km orbit started in May 2000. We remained in a steady orbit 50 km from the center of mass of the asteroid until, in July 2000, we went into a two-week period at 35 km and then returned to 50 km before going back into another set of high orbits.

The entire low-orbit period was a time to recuperate from the grind of orbit insertion and high orbit. The XGRS team needed to view each part of the asteroid for long periods, so they usually pointed NEAR Shoemaker to a fixed position and let the asteroid rotate under them for most of the day with no fancy mosaics or movements of the spacecraft. MSI and NIS would continually shoot pictures and spectra as the asteroid rotated, which provided imaging for the XGRS team and entirely covered the asteroid at this resolution. The MOPS and MSI teams welcomed the relief of the predictable routine. But the NAV team had a different problem. The error accumulated during the five-week low-orbit operational phase meant that NAV needed to update the on-board trajectory more often to keep the instrument pointing errors to a minimum.

The 50-km orbit was retrograde; that is, NEAR Shoemaker was moving one way while Eros rotated in the opposite direction. This kind of orbit is preferred,

because the gravitational "tugs" from the ends of Eros do not last as long if the spacecraft and Eros are moving in opposite directions. In the opposite scenario, a so-called "direct" orbit, the spacecraft would spend much more time over each end, and the orbit would change much more quickly. Before too long NEAR Shoemaker would have either crashed on to Eros or been thrown completely out of orbit. For this reason we had no choice but to have NEAR Shoemaker travel in a retrograde orbit. However, with the retrograde orbit and the direction of the Sun, the spacecraft was forcing itself into an "uncomfortable" position and building up momentum within its reaction-wheel pointing system whenever it pointed to the asteroid; it needed to unspin these wheels weekly through small thruster firings to release or "dump" the unwanted momentum.

Every day during the low orbits the imaging team had to point three opnavs for the NAV team. The opnavs now consisted of pairs of observations. Bill Owen identified 44 craters on the surface for MSI to target as often as possible. In order to get the best possible viewing of a crater it needed to be almost directly under the spacecraft and close to the terminator. By making observations a few minutes of time apart of a pair of craters separated by several degrees on the surface of the asteroid, the NAV team could pinpoint NEAR Shoemaker's location in three dimensions to a few meters. The navigators also used 50% of the observations taken during the fixed pointing time period. This gave NEAR Shoemaker the most accurate pointing possible and led to Bill Owen (as well as many of the science team members) knowing almost every small crater on the surface of the asteroid intimately.

The targeting of the opnavs proved to be a tedious job for MSI so Brian Carcich set out to make our lives easier once again. In his *orbit* program, he made it possible for us to place a cursor on the asteroid and simply click on a desired feature to save the x, y, and z coordinates to be put into the pointing command. He also created a process by which we could easily identify and target opportunities for imaging of any of the craters that were on the NAV team's list of favorites. It was ingenious and made our lives bearable for the week-to-week grind of targeting and turning in sequences.

During the 50-km orbit, discussion began about the possibility of a low-altitude flyover (LAF), moving the spacecraft for a short time very close to the asteroid. The NIS team had originally planned a second low-phase flyby in October but, because of the instrument's untimely demise in May, that period opened up as a good opportunity to try out the LAF idea. At the same time, planning of the final phase of the mission, when the spacecraft would run out of fuel, began. The idea of a landing on the asteroid had been kicked around for several years. No one had ever done it before so it was intriguing. All of these ideas landed on the shoulders of the NAV team. Could they put together the sequence of thruster firing events that needed to happen to have a low-altitude flyover and have the spacecraft survive?

The 35-km orbit in July provided additional tests of the maneuverability of the spacecraft, as well as the best opportunity to determine the gravity field of Eros from radio science measurements. The XGRS team found that their gamma-ray instrument was not as sensitive as they had hoped and they needed to get closer to capture the weak gamma radiation. NAV considered the opportunity to be perfect for testing their ability to maneuver the spacecraft closer to the asteroid. The 35-km orbit was a success from the point of view of the navigation team. They had come in close, orbited for a week and pulled out to 50 km without any problems. Unfortunately for the gamma-ray team, there was a solar flare that was so intense that it shut down the instrument temporarily. XGRS had lost valuable data but NAV had gained valuable experience.

Flyovers and the landing

In September of 2000, we were back in high orbit. The NAV team sent a LAF test trajectory to the science teams for their approval. They had found a way to do a low-altitude flyover that involved a number of thruster firings to bring the spacecraft close (from 100 km to 50 km to 35 km to 20 km). The last burn would sling shot the spacecraft into a highly elliptical orbit that came within 5 km of the asteroid's surface. The MSI team worked for most of September and October to create a set of commands that would maximize the viewing of the asteroid while taking into account the possible errors in the trajectory. We did not want to end up taking pictures of unlit asteroid or black space just because the trajectory was off by a degree or two.

Rick Shelton on the MOPS team and Maureen Bell on the MSI team stayed in daily contact during most of October, testing and tweaking the commands to get the best possible product. Everything went smoothly until three days before the LAF burn. Then we received a call from Karl Whittenburg in MOPS. Murphy's Law was with us and the spacecraft had gone into what is called "safe mode." NEAR Shoemaker's on-board computer was programmed to monitor the spacecraft's position and health constantly. When the computer software recognized an unacceptable situation, NEAR Shoemaker would automatically turn itself into a safe position with its antenna pointed toward Earth asking for help and instructions. It was quickly determined that the cause was minor rather than a spacecraft health problem, so the LAF burn was still on but without the optical observations for the days leading up to it. The Doppler data were vital to the NAV team for determining the pre-burn trajectory.

On October 25, 2000, NEAR Shoemaker swung in towards Eros, took its pictures and darted back out into a 200 km orbit. The data collected during the close encounter were played back three times to be sure that nothing was missed. It was a highly successful encounter. The navigation again worked perfectly. The spacecraft thrusters performed flawlessly and the images were all of the illuminated asteroid. The science team was thrilled with the diversity of geology studied for the first time at this scale on such a small solar system body. Now all the teams were ready for the final set of navigation challenges.

Another low-altitude flyby was planned for January 2001. This one would bring NEAR Shoemaker in even closer to the asteroid (within about 2 km) and it would maintain an elliptical orbit around 20 km. It was several days long and included several downlink periods and close flyby periods. It was a new level of difficulty for the NAV, MOPS, and MSI teams. This time Murphy left us alone. The thruster firings

Figure 8.10. The predicted footprints of images taken as NEAR Shoemaker attempted its first low-altitude flyover on October 26, 2000 (left). The image mosaic on the right is a small portion of the images taken just after the closest approach, which was between the two craters that are shadowed in the footprint plot.

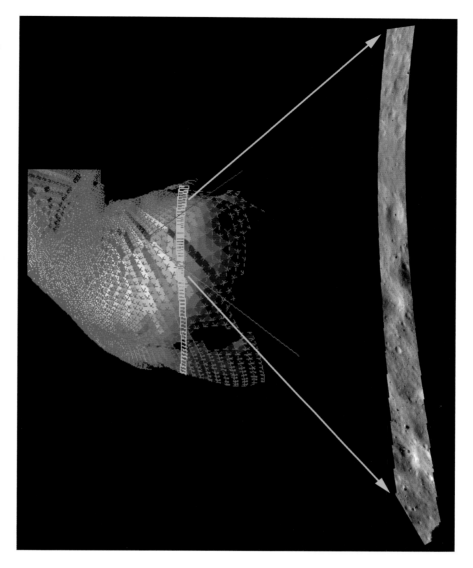

occurred without incident, and the pictures were even more spectacular.

The final challenge was landing on the asteroid. The process of determining where and how to land had started back in the summer of 2000. In November, at a meeting of the NAV, MOPS, and MSI teams, it became obvious that the descent trajectory plus the pointing constraints determined our landing site for us. In order to take pictures of the asteroid as we descended to the surface, it was necessary to maintain a very specific configuration. The spacecraft's high-gain antenna, which transmitted data to Earth, had to remain pointed to Earth constantly during the landing. The solar panels had to stay pointed at the Sun to maintain power. The cameras had to point at the surface of the asteroid. The antenna, panel, and camera are all at right angles to each other so there was only one small subset of options if we wanted to be in this configura-

tion. The constraints led us to a landing at latitude 35° South. The position and trajectory were fixed. NAV designed a series of thruster burns to bring the spacecraft in closer. Karl Whittenburg created special commands for the spacecraft for each stage of the landing. He and Ann Harch of the MSI team worked out a set of camera commands that would take images continuously and transmit them to Earth in pairs. (NEAR Shoemaker stored all images on to a solid state data recorder and then transmitted them to Earth; there was no option for direct transmission of data.)

All the teams converged on APL on February 12, 2001, except for the NAV team who still had the job of tracking the spacecraft back at JPL. We had all worked hard for a year and felt pretty happy with the results and with the very cool way that the spacecraft was ending its life. No one expected the spacecraft to survive. The mission operations center was packed

Figure 8.11. The location of the landing site for NEAR Shoemaker's "controlled descent" to the surface on February 12, 2001. The main image shows the touchdown site (yellow circle) on the edge of the saddle-shaped feature Himeros. The inset is a mosaic of eight images showing the site in the context of the eastern part of the southern hemisphere of Eros. The landing site straddles two major terrain types on Eros: older, heavily cratered southern highlands, and the younger, less heavily cratered interior of Himeros.

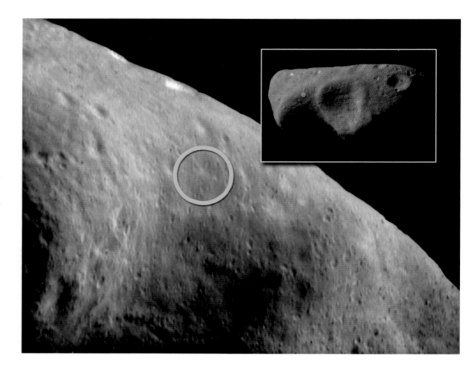

Figure 8.12. The predicted footprints of images taken as NEAR Shoemaker descended to the surface of Eros on February 12, 2001 (bottom). During the descent, NEAR Shoemaker took a strip of more than 70 images, crossing from the southern highlands into Himeros. The image mosaic of 15 of the final images taken by NEAR Shoemaker's cameras are shown at the top. The final images show details on the surface smaller than 10 cm (4 inches) across.

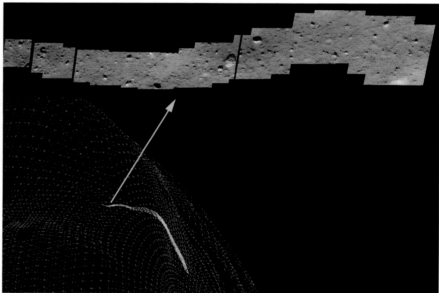

with the NEAR management team, the operations people, and a few press people and dignitaries. Mostly it was a room filled with anticipation; everyone was waiting to see whether NEAR Shoemaker could really do this spectacular feat.

The descent thruster firings began, the spacecraft descended, and images began to roll in. In the image-processing room at APL, we saw the pictures first and processed them, then sent them on to the mission operations center to be displayed on screen for the television cameras. We all sat together reading off the values of the height from which each image had been taken, marveling that everything was working perfectly. When the final picture came in we were startled to find out that the spacecraft had "landed" two minutes earlier than expected and was performing a burn during the landing. But even this contingency had been planned for – the onboard accelerometers detected the jolt, determined that we had supplied enough downward thrust, and then most likely reversed the thrusters. This had created a soft landing. Otherwise we might have lifted back off from the

Figure 8.13. The media and science teams converged on the JHU APL Mission Operations Center (top) and the JPL navigation office (the war room, bottom) to watch NEAR Shoemaker's historic descent to the surface of Eros. Top photo: APL MOPS team members from left to right: Owen Dudley, Carolyn Chura, Robert Bokulic, Karl Whittenburg, and Robert Nelson. Bottom photo: JPL NAV team members from left to right: Margie Medina, Steve Chesley, Cliff Helfrich (head partly hidden), Mike Watkins, NAV team leader Bobby Williams (holding pen), and Pete Antreasian (foreground).

surface, or at least bounced. The spacecraft survived and maintained its antenna alignment with Earth (miraculously) and began sending back its navigation information.

Events unfolded with similar excitement at JPL. The NAV "war room" was electric, with the team still working hard. NAV team leader Bobby Williams was on the special voice line to APL; Pete Antreasian, Steve Chesley, and Eric Carranza were at computers, monitoring the real-time data; Jim Miller was watching the predicted descent curve and calling out numbers; Mike Wang and Bill Owen were running back and forth between the war room and the opnav room, where they would quickly download the latest

images. There was a closed-circuit video feed from MOPS in one corner of the room. Maybe 25 other folks squeezed into the room, mostly navigators on other projects, but also a news crew. One computer monitor showed the altitude measurements superimposed upon the predicted distance from NEAR Shoemaker to the ground. The measurements fell right on top of the line, indicating that all the maneuvers had done their jobs and we were on the right course. We all realized that the events we were watching had already happened some 17½ minutes before, and that we were powerless to change things in any event, but this knowledge did nothing to dispel the tension. Then the Doppler data changed abruptly, and four of the six DSN antennas that were listening to NEAR Shoemaker lost lock. But two Goldstone antennas continued to receive a signal! Not only had we landed, but NEAR Shoemaker was still alive and well! Nobody expected that. It was a phenomenal end to a year of driving a spacecraft.

Although it would have been fun to fire the engines again and bump the spacecraft off the surface, there was not enough fuel left. So NEAR Shoemaker performed one more task. It took gamma-ray data for 12 days and transmitted these back to Earth. It was put to sleep in a safe configuration so that the team could try at some future date to do something that had never been done before – awaken a computer that has slept through a long cold night on the surface of an asteroid.

The NAV team had been planning for the NEAR mission and landing for several years. At a conference in August 1999 they had presented a paper called "Preliminary Planning for NEAR's Low-Altitude Operations at 433 Eros." In the abstract they had written, "This paper will provide preliminary plans for mission design and navigation during the last five weeks of the orbit phase, where several close passes to the surface will be incorporated to enhance the science return. The culmination of these close passes will result in the eventual impact of the spacecraft on the surface of Eros. The possibility of hovering within 1 km from Eros's surface exists and could be incorporated into a landing design." The JPL navigation team accomplished more than they dared hope for, with the help of the teams at APL and Cornell. They managed to keep the spacecraft on track so that it could take a total of almost 200 000 MSI images, and similarly large amounts of data for the other science instruments. Despite many near-tragedies, it was, in the end, a picture perfect mission.

9 Mission Accomplished

Andrew Cheng

Applied Physics Laboratory, Johns Hopkins University (NEAR Project Scientist)

The curtain came down on NEAR Shoemaker's adventures at Eros on February 28, 2001, when the final message from Earth silenced all communication from the spacecraft. It now nestles where it gently touched down within the large crater Himeros. Its soft landing was a spectacular encore to its acclaimed performance. There had been a moment of high drama followed by elation when the mission teetered back from the brink of disaster. But, by the end of the show, the plucky little spacecraft in the hands of its dedicated team of scientists and engineers had thrilled the world, deftly executing every trick demanded of it.

Not only was the NEAR mission a great success scientifically, it also earned a special place in space history by attaining several "firsts". As the first space mission dedicated to studying the geophysics, geochemistry, and geology of asteroids, it made the first spacecraft flyby of a C-type asteroid, 253 Mathilde, and then became the first to orbit an asteroid – 433 Eros. As if that was not enough, its landing on Eros and the return of scientific measurements from Eros's surface were both further spacecraft firsts.

NEAR was truly a remarkable mission. Now, several months after the final data were returned, we can stand back and take an overview of its achievements, the discoveries it has made, and the many new questions posed by its revelations.

NEAR in a nutshell

NEAR's main objective was to carry out a year-long comprehensive scientific study of one of the near Earth asteroids (NEAs) – asteroids that come within 1.3 AU of the Sun. Eros was chosen as the prime target not only because of its large size and accessible orbit, but also because of its scientific importance. The maximum dimension of Eros's elongated shape is 32.7 km, making it the third largest NEA known. It

was the first asteroid to be discovered in an orbit that brings it closer to the Sun than Mars, and the first asteroid found to vary in brightness as it rotates. (The brightness changes arise because the illuminated surface area of the irregularly shaped asteroid varies as seen from Earth.) More recently, Eros has been a focus of debates regarding the possible relationship between S-type asteroids and ordinary chondrite meteorites, and studies of its chaotic orbit have shown that it could someday in the far future collide with Earth, although it is more likely to plunge into the Sun or be ejected from the solar system altogether.

The three-axis-stabilized NEAR Shoemaker spacecraft carried a five-instrument scientific payload, composed of a multispectral imager (MSI), a near-infrared spectrometer (NIS), an X-ray/gamma ray spectrometer, a laser rangefinder, a magnetometer, and a radio science investigation using the spacecraft's

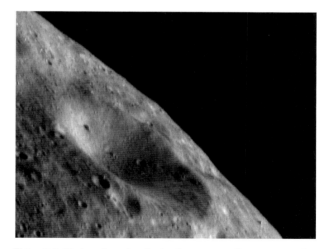

Figure 9.1. Stark and spectacular landscapes were the norm once the NEAR spacecraft began routine operations in orbit around Eros. In this image, taken April 17, 2000, from a height of 101 km (63 miles), the surface is pockmarked with craters. The one in the centre is 2.8 km (1.74 mile) in diameter. The smallest craters and boulders that can be resolved are about 20 m (65 feet) across.

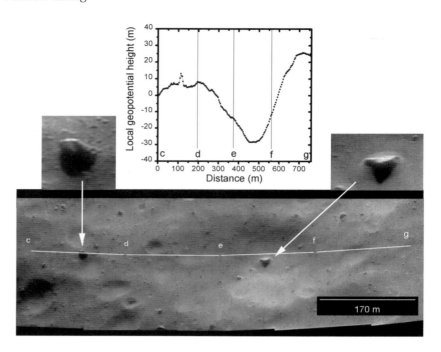

Figure 9.2. Among the many "firsts" of the NEAR mission were the first laser altimetry profiles of an asteroid. An example of a NEAR Laser Rangefinder (NLR) topographic profile is shown here, obtained from a range of 6.5 km (4 miles) on October 26, 2000. The image mosaic at the bottom was obtained simultaneously with the laser data. The white line on the mosaic marks the laser track, and corresponding points on the image mosaic and in the graph of geopotential height, or elevation, are marked with letters c through g. The laser sampled a boulder between c and d (small left inset). The laser sampled a "ponded deposit" between letters e and f. The boulder in the pond is shown in the small right inset. Many small raised features in both insets appear elongated and roughly aligned.

coherent X-band telemetry system. NEAR's scientific program studied the asteroid's surface morphology, surface composition, and interior structure.

The first science return came on June 27, 1997, when NEAR Shoemaker executed a flyby of the main-belt asteroid 253 Mathilde. Subsequently, NEAR Shoemaker flew by Earth again on January 23, 1998, receiving a gravity assist which targeted it towards Eros. On December 20, 1998, NEAR Shoemaker was scheduled to begin its rendezvous with Eros, but the first engine burn unexpectedly aborted, and contact with the spacecraft was lost for 27 hours. After recovery of communications, NEAR Shoemaker flew past Eros on December 23, 1998. A successful maneuver on January 3, 1999 set NEAR Shoemaker on course for a return to Eros in February 2000. The 1998 Eros flyby yielded important first measurements of the mass and shape of Eros, which in due course reduced the risk attached to initial orbital operations at Eros.

On 14 February 2000, NEAR Shoemaker passed directly between the Sun and Eros, and the highest priority near-infrared spectral maps were obtained successfully at low-phase angles (when shadows on the surface are minimized). Later that same day, the spacecraft's insertion into orbit around Eros was accomplished successfully. From then until 12 February 2001, NEAR Shoemaker performed orbital operations, including global mapping orbits at 100 km, 200 km and higher (measured from the center of Eros), geochemical mapping orbits at 50 km and lower, and laser ranging orbits at 50 km. In general, all

instruments acquired data simultaneously, but a particular investigation had priority at any one time for specifying instrument pointing. In the geochemical orbits, for example, the X-ray/gamma ray team had priority for pointing. On October 26, 2000 NEAR Shoemaker executed a low-altitude flyover at a minimum altitude of 5.3 km from the surface. More low-altitude flyovers took place during the last week of January 2001 in preparation for the landing on 12 February 2001. All instruments operated nominally throughout the mission with the exception of the NIS, which failed on May 13, 2000. The NEAR rendezvous with Eros obtained the first X-ray and gamma ray spectra and the first laser-ranging measurements of an asteroid.

NEAR in context

Asteroids are part of a vast population of small bodies orbiting the Sun in addition to the nine major planets. The largest of the small bodies is asteroid 1 Ceres with a diameter of 935 km (584 miles). It is smaller than the smallest of the planets, Pluto, whose diameter is 2400 km (1498 miles). The small bodies are often described as primitive bodies because many of them have remained virtually unchanged since the formation of the solar system some 4.6 billion years ago. They thus preserve a record of physical and chemical conditions in the early solar system. On Earth, ancient rocks more than 4 billion years old have been destroyed and cannot be found on the surface, except for the

Figure 9.3. On February 12, 2001, the NEAR Shoemaker spacecraft came to rest gently along the slopes of the large crater/depression Himeros. The best estimate of the actual landing site is shown here, along with the fields of view of the final few images acquired as the spacecraft descended (yellow boxes). These images are discussed in detail in Chapters 4 and 7.

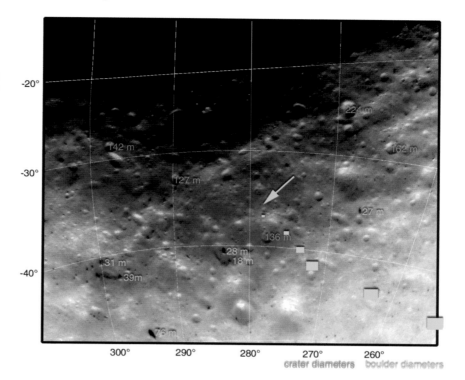

extraterrestrial rocks we call meteorites. Meteorites are fragments of asteroids that fall to Earth from deep space. It is partly from laboratory analyses of meteorites that the age of the solar system has been determined.

If a small body has retained enough ice (water ice and other frozen materials) to create a detectable coma or tail when it nears the Sun, it is a comet; otherwise it is an asteroid. Some asteroids are presumably comets that have lost their ice. "Meteoroid" is a term often used for objects smaller than asteroids or comets. There is no agreed figure for the maximum size of a meteoroid but, in current usage, an object would be considered to be a meteoroid if it is too small, and therefore too faint, to be observable even with the aid of a telescope. Such objects are usually less than about 10 m (33 feet) across.

The most primitive asteroids and comets are remnants of the original small bodies – planetesimals – that coalesced to form the planets. Other asteroids are fragments of objects that underwent partial or complete melting and evolved much of the way to becoming planets in their own right, but were destroyed by catastrophic collisions early in the history of the solar system. For instance, some asteroids are composed of high density materials like those found in the core of the Earth. The small bodies in the solar system include not only the original building blocks of planets, but

Figure 9.4. One of the major surprises from NEAR's imaging investigation was the lack of color variations on the surface associated with different geologic terrains, despite the sometimes large brightness differences between these terrains. For example, NEAR Shoemaker took these images of Eros on October 16, 2000, while orbiting 54 km (34 miles) above the asteroid. The top panorama shows the rounded rim of the saddle-shaped feature Himeros. Parts of the plains surrounding Himeros (lower left) and inner wall of Himeros (lower right) are also shown. These are false color composites, constructed from images taken in green light and two different wavelengths of infrared light. Surface materials that have been darkened and reddened by the solar wind and micrometeorite impacts appear as pale brown, whereas fresher materials exposed from the subsurface on steep slopes appear in bright whites or blues. Compared with Gaspra and Ida, similar asteroids imaged in color from the Galileo spacecraft, Eros exhibits large brightness variations but only subtle color variations.

also samples of the exteriors *and interiors* of potential planets that were broken to bits while still in the process of forming. The small bodies as a group are as diverse as the planets themselves.

Different types of asteroids are classified according to the way they reflect sunlight in the visible part of the spectrum. More than a dozen asteroid spectral types are generally recognized, including the S-type (stony) asteroids, the C-type (carbonaceous) asteroids, the M-type asteroids (some of which are metallic and some hydrated), and the D-type asteroids, which are primitive and rich in organic (carbon-bearing) minerals like the C types, but redder. The S-type asteroids are the most numerous type in the inner solar system. These objects contain the silicate minerals pyroxene and olivine as well as metallic iron.

The C-type asteroids dominate the central portion of the main belt of asteroids between Mars and Jupiter. Their dark, neutral spectra suggest a carbonaceous composition, based on their similarity to the spectra of carbonaceous chondrite meteorites. The carbonaceous chondrites come from parent bodies – presumably asteroids – that have undergone varying degrees of alteration by water and heat. The histories and origins of these primitive meteorites, and the question of whether they are related to primitive asteroids (including the C-types), to the dark materials found in comets, to trans-Neptunian or Kuiper Belt objects, or to the satellites of the outer planets, are among the most important unresolved issues in solar system exploration.

NEAR Shoemaker visited two distinctly different types of asteroids. Its encounter with Mathilde provided the first close-up look at a C asteroid. Then the surveillance of Eros from orbit, and the landing on its surface, were the first comprehensive exploration of an S-type asteroid. The two asteroids visited by the Galileo spacecraft, 951Gaspra and 243 Ida, were also S asteroids.

The NEAR mission addressed two fundamental questions about asteroids. The first is whether some S-type asteroids are examples of primitive asteroids that have not melted or subsequently "differentiated" into bodies with distinct core–mantle–crust layered structures. There are various subtypes of S asteroids, some of which are believed to be differentiated (including Gaspra, based on the Galileo flyby data). However, the subtype to which both Eros and Ida belong was suspected from telescopic observations to be primitive. A primary objective of the NEAR mission was to determine whether Eros, and by inference similar S asteroids, are primitive or differentiated objects. To do

Figure 9.5. The discovery of mottled bright and dark terrains like these was another surprise from the NEAR mission. This image of the interior of Himeros was taken on May 9, 2000 from an orbital altitude of 49 km (30 miles). In addition to the bright and dark markings that range from 20 to 400 m (65 to 1300 feet) across, boulders can be recognized down to the limit of the resolution – about 8 m (26 feet). The whole scene is about 1.8 km (1.1 miles) across.

this, the NEAR Shoemaker X-ray and gamma-ray spectrometers measured the abundances of key elements such as silicon, magnesium, and iron. In addition, visible and near infrared spectra taken by NEAR Shoemaker were used to infer what silicate minerals are present on the surface and it searched for any intrinsic magnetic field, which would give hints as to Eros's thermal history.

A related issue is the link between asteroids and meteorites. While laboratory measurements of meteorites provide detailed information on their physical and chemical properties, compositions, ages, and thermal histories, for example, it has remained a challenge to infer the properties – or even the identities – of meteorite parent bodies in the asteroid belt. It is natural to expect that the most common meteorites (ordinary chondrites) should originate from the most common asteroid type (S) in the inner asteroid belt, but the spectra do not match. Is Eros related to the ordinary chondrites? In general, matches are hard to find between asteroid and meteorite spectra aside from the known association between the basaltic HED meteorites and the uncommon type V asteroids, like 4 Vesta. This puzzle has led to conjectures about whether so-called "space-weathering" processes have altered the surfaces of asteroids in the asteroid belt.

The second fundamental question deals with the collisions that took place between small bodies in the early solar system, in the era when the terrestrial planets formed. A key issue is the balance between the violent impacts that tore bodies apart and the accretion of small bodies to form larger ones as a result of

Figure 9.6. A variety of geologic processes appear to be at work within and surrounding Eros's largest crater, Psyche. For example, several troughs and scarps appear to cut through the crater from lower left to upper right. Bright and dark markings on the crater walls probably come from dark material moving downslope and revealing fresher material underneath. And several large boulders, probably ejected from this or other cratering events, perch precariously on the crater walls, seeming to defy gravity on this small, irregular body. This mosaic was acquired on September 10, 2000 from an altitude of 100 km (62 miles). The crater is approximately 5.5 km (3.4 miles) in diameter.

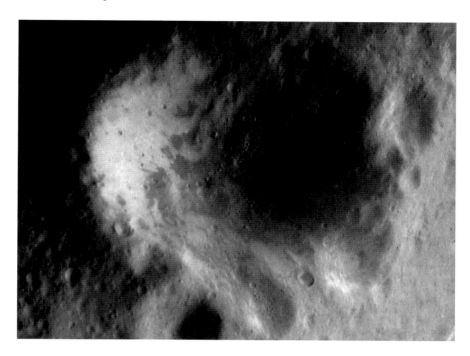

gentler collisions. The question is whether Eros is a so-called "rubble pile" – an agglomeration of smaller bodies bound together by gravity, an intact fragment from a larger parent body, or something in between.

NEAR and Mathilde

NEAR Shoemaker's encounter with Mathilde was highly significant. It was the first close-up look at an asteroid that is completely different from the S types already explored by Galileo and NEAR's main target, Eros. As a C asteroid, Mathilde is of the type that predominates in the central portion of the main asteroid belt. Apart from its importance as the first example of a C asteroid to be explored, Mathilde is interesting because it rotates extremely slowly. Its 17.4-day period is the third longest known and is at least ten times longer than that of typical asteroids. Why some asteroids rotate so slowly is still puzzling. Mathilde certainly does not have any satellite large enough to explain plausibly the slow rotation as a result of tidal forces. Despite a search by NEAR, no natural satellite of Mathilde was found, although a few main belt asteroids including Ida are known to have satellites.

The NEAR Shoemaker flyby of Mathilde obtained the first direct mass determination of an asteroid. The measured mass of 1.03×10^{20} g and estimated volume of 78 000 km^3 imply a density of 1.3 ± 0.3 g cm^{-3}. Mathilde's volume had to be estimated because only one face of it could be imaged during the 25-minute flyby. The inferred density was unexpectedly low at

half, or even less than half, of that of carbonaceous chondrite meteorites, which are the closest analogs on the basis of their spectra. It implies that Mathilde is not solid, but porous, with a high porosity of 50% or more. The surface of Mathilde is heavily cratered. There are at least five giant craters with diameters comparable to the 26.5-km mean radius of the asteroid. Such enormous craters are about the largest that could be created without destroying Mathilde. It is remarkable not only that Mathilde survived at least five giant impacts, but also that the latest of the impacts did not destroy the previous craters. The likely explanation is that Mathilde's porous nature helps dissipate the energy of an impact, reducing its destructive power. The porous material acts like a shock absorber, like pounding a bag of sand with a hammer. It is easily compacted and its voids may interfere with the propagation of shock waves. Finally, Mathilde proved to be remarkably uniform in both color and albedo. Previous telescopic observations could not rule out the possibility of small bright patches (of ice, for example) or spectrally distinct regions. The NEAR Shoemaker observations revealed no evidence of any albedo or spectral variations, implying that Mathilde's composition is the same throughout.

Mathilde's porosity may be microscopic, macroscopic, or both. It could be how Mathilde has always been or the result of a catastrophic event. The asteroid could have accreted as a porous structure and survived as such to the present. This picture would suggest microscopic porosity similar to that of interplanetary

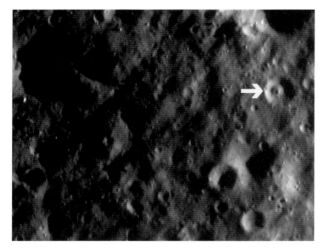

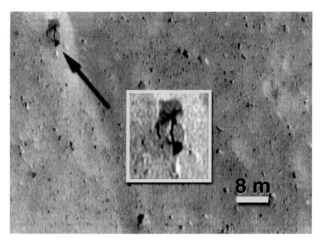

Figure 9.7. High-resolution global imaging from NEAR continually revealed new and interesting classes of features on Eros during the course of the orbital mission. An example is shown in this image taken on June 14, 2000, from an orbital altitude of 50 km (31 miles). At first glance it looks like a region of typical heavily cratered terrain; however, the 200-m (656-foot) diameter crater at the arrow is unusual because of the sizeable mound of material on its floor. Mounds inside craters can originate in various ways, sometimes during the process of crater formation, and sometimes during later modification of the crater by surface processes. This type of feature has been seen previously on larger planetary bodies like the Moon and Mars, but never on a small, low-gravity object such as Eros.

Figure 9.8. Images like this one of the surface of Eros at extremely high resolution provide evidence that there is probably a substantial layer of shattered, fragmented debris (regolith) accumulated globally across the asteroid. In this image, a large boulder (inset) appears to have broken into several smaller pieces, probably as it impacted the surface after being ejected by a cratering event elsewhere on the surface. In this image, acquired as the spacecraft was in its final descent to the surface on February 12, 2001, rocks and other features as small as a few tens of centimeters can be resolved.

dust particles. Alternatively, the structure may be a gravitationally bound agglomeration of boulders formed when a thoroughly fractured Mathilde was disrupted by impacts but not dispersed, so that it subsequently reassembled. In this case, macroscopic voids would be expected, possibly in addition to microscopic porosity. Another possibility is that fragments of diverse bodies were accreted to form Mathilde. However, there is no evidence of any layered structure or of any variations in its composition, despite the presence of giant craters that have excavated the surface and exposed material kilometers below the surface. If Mathilde is made of accreted fragments of diverse parent bodies, they must have had remarkably uniform albedos and colors, or else the fragments must be smaller than about 500 m and impossible to resolve in the NEAR images.

NEAR at Eros

The first results from Eros's orbit have shown that Eros is a primitive, undifferentiated asteroid and a consolidated object. The bulk elemental composition of Eros is consistent with that of ordinary chondrites based upon the areas so far analyzed, but a primitive achondrite-like composition is not ruled out. The main

difference is that Eros has less sulfur. The silicate mineralogy of Eros, inferred from visible and near-infrared spectra, is consistent with low-iron ordinary chondrites. Eros has neither melted nor differentiated fully, but some degree of partial melting or differentiation is possible. No evidence for intrinsic magnetization of Eros has yet been found, even during the final descent. If Eros were as magnetized as typical meteorites, the magnetization would have been easily measurable. The absence of magnetization may be consistent with a thermal history in which Eros was never heated to its melting point. There are subtle variations in spectral properties across the surface, but no firm evidence for variation in composition has been found.

The average density of Eros, as first found during NEAR Shoemaker's December 1998 flyby and since confirmed in orbit, is about the same as that of Earth's crust. This average density of 2.7 g cm^{-3} is less than the average bulk density of ordinary chondrite meteorites as measured in the laboratory, suggesting that bulk Eros is significantly porous and/or fractured, but not to the same extent as Mathilde. The density of Eros's interior is nearly uniform; its gravity field is within a few percent of what would be expected of an object with Eros's shape that is the same density throughout. There is a small offset between the center of mass and the center of figure, which may be consistent with a regolith layer up to 100 m deep in places.

NEAR Shoemaker has shown that Eros is a

Figure 9.10. There is an amazing amount of evidence for the motion of regolith materials across the surface of Eros in high-resolution NEAR images such as this. The upper half and lower right parts of the image show surfaces with "typical" rounded craters and large boulders. However, the sharp-edged streaked terrain extending from lower left to middle right appears remarkably more smooth, subdued, and lacking in small-scale detail of any type, almost as if Eros had been altered by a giant eraser. This image was obtained on January 7, 2001 from an orbital altitude of 35 km (22 miles). The whole scene is about 1.4 km (0.9 miles) across.

Figure 9.9. The variety of geologic features that NEAR found on Eros was surprising to many astronomers who had assumed that the asteroid's small size and low gravity would preclude the existence of any "interesting" geology on the surface. For example, this picture, taken July 24, 2000 from an orbital altitude of 36 km (22 miles), shows a region about 900 m (3000 feet) across that contains craters, fragmented and partially buried boulders, and "ponded" deposits of loose regolith materials that appear to be filling in topographic low spots.

consolidated body and not a "rubble pile." A variety of ridges, grooves, and chains of pits or craters pervade its entire landscape. Coherent systems of linear features extend globally. The shapes of many craters appear to have been influenced by existing structure in the fabric of Eros. There are slopes too steep to be mounds of loose material and so indicate the presence of a consolidated substrate. Tectonic features include one ridge that extends over 15 km across the surface. These findings all suggest that Eros is a fragment from a larger, undifferentiated, parent body. And it seems to be alone, as the presence of a satellite larger than 20 m in diameter has been ruled out.

Most of Eros's surface is old and close to being saturated with craters, but some regions appear to be

relatively young and extensively resurfaced. Blocks and boulders are ubiquitous and are not confined to gravitational lows. The surface is extremely rough and exhibits a fractal structure on scales from a few meters up to hundreds of meters. Examples have been found where material has slipped down steep slopes in crater walls. Bright areas are also associated with steep crater walls. The morphological and spectral properties of these striking features are consistent with a scenario in which darkened, space-weathered material slides down crater walls exposing brighter, less-weathered underlying material.

The discovery of extremely level deposits, dubbed "ponds" was both striking and unexpected. They were seen in the high-resolution, low-altitude observations made toward the end of the mission. Processes such as electrostatic levitation and seismic shaking from impacts have been discussed as possible means to mobilize fine particulates on the surface. Finally, after landing on the surface, NEAR obtained its best-quality gamma-ray measurements of the surface composition. These data will help to resolve the question of how much thermal metamorphism Eros has undergone.

Outstanding issues

No one should be surprised that the NEAR mission has raised as many – if not more – questions as it has

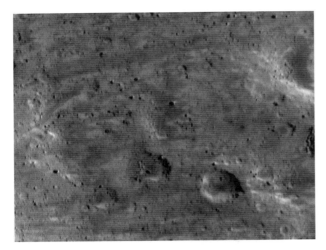

Figure 9.11. One of many examples of striking and surprising geologic terrain discovered in NEAR imaging data. This image of the southern part of Himeros was acquired on June 17, 2000, from an altitude of 51 km (32 miles). Small craters are mostly absent and, in larger craters, the rims appear highly degraded and the floors flattened. These morphologic characteristics, when they appear on larger planets, typically are interpreted as evidence of blanketing of an older surface. That surface modification processes of this kind could also be at work on such a small object as Eros was a surprise to many astronomers. The whole scene is approximately 1.9 km (1.2 miles) across.

Figure 9.12. Could this be the origin of the enigmatic asteroid 433 Eros? Eros was imaged by the MSI camera just after orbit insertion on February 14, 2000, and the resemblance of the asteroid in this particular orientation to the ancient "moai" stone statues of Easter Island was uncanny. While NEAR scientists had some good-natured fun with this cosmic coincidence, it will likely take years of detailed study of the images and other measurements obtained by NEAR Shoemaker to discover the real origin of Eros. Fanciful image processing by Elaina McCartney, Zvi Shiller, and Phil Chu.

answered. Every time a spacecraft surveys one of the worlds of the solar system for the first time, be it large or small, there are surprises and puzzles. No two worlds have ever been found to be alike. NEAR Shoemaker has revealed to us in very considerable detail what kind of object Eros is, even though uncertainty about the nature of its interior remains. The main challenge now is to understand "why."

Here are some questions that remain to be resolved:

• Is Eros related to a known meteorite type, and if so which? The visible and near-infrared spectra are consistent with low-iron ordinary chondrite meteorites, but why is sulfur depleted?
• The darkening and reddening of Eros's surface produced by space weathering are different from the effects of space weathering on the Moon. Why?
• Eros has albedo variations of up to a factor of two, but Ida shows no such variations. However, Ida has 20% visible color variations, but Eros has virtually none. Why are Eros and Ida so different?
• Some color variations on Ida are associated with crater ejecta. Why is there so little evidence of ejecta on Eros – unless the ubiquitous boulders are just that?
• Why are there so few craters on Eros with diameters measuring 100 m or less? Why are there so many boulders, especially ones tens of meters across and smaller?
• How deep is the regolith on Eros and how is it distributed? Globally, regolith depth is at most of the order of 100 m; in some places it appears to be about this deep, but only meters thick in others. Why?
• What makes "ponds"? Small collapse features in the regolith? Debris aprons around some boulders? Electrostatic levitation? Seismic shaking?
• Is Shoemaker Regio the most recent large impact on Eros? Is Himeros an impact crater? Where are the boulders ejected from Psyche crater?
• How consolidated, and how fractured, is Eros? How much tensile strength does Eros have? Do Rahe Dorsum and its associated features indicate at least one fracture of global scale?
• Eros is about 20% void space, but where and how large are the pores or fractures? Are they microscopic or macroscopic?
• Is Eros tumbling? If so, the amplitude is less than 0.1 degree.
• Is the interior of Eros inhomogeneous?
• Why is Eros so non-magnetic?

If there is one lesson to be learned from NEAR, it is that asteroids are varied and surprising. Although Eros has been studied in ways never attempted for any other asteroid, and it has been imaged at resolutions of a few centimeters, it remains mysterious in many ways.

Bibliography

BOOKS

Binzel, R. *et al.* (eds) *Asteroids II*. Tucson, AZ: University of Arizona Press, 1989.

Gehrels, T. (ed.) *Asteroids*. Tucson, AZ: University of Arizon Press, 1979.

Kowal, C. T. *Asteroids: Their Nature and Utilization*, 2nd edn. New York: John Wiley & Sons, 1996.

McSween, Harry Y., Jr. *Meteorites and their Parent Planets*, 2nd edn. Cambridge University Press, 1999.

Norton, O. R. *Rocks from Space*. Missoula, MT: Mountain Press Publishing, 1994.

Russell, C. T. (ed.), *The Near Earth Asteroid Rendezvous Mission*. Kluwer Academic Publishers, 1997.

GENERAL ARTICLES

Beatty, J. Kelly NEAR falls for Eros. *Sky and Telescope*. 101(5), May 2001, 34–37.

Bell, J. F. Far journey to a NEAR asteroid. *Astronomy*. March 1996, 42–47.

Chapman, C. R. Small worlds, up close. In *Our Worlds: The Magnetism and Thrill of Planetary Exploration*, ed. S. A. Stern, 65–91. Cambridge University Press, 1999.

Chapman, C. R. Asteroids: up close and personal. *Physics World*, June 2001, 33–37.

Farquhar, R. *et al.* The Second Coming of NEAR. *The Planetary Report*, November–December 1999, 14–18.

COLLECTIONS OF TECHNICAL PAPERS

NEAR spacecraft and mission design. *Journal of the Astronautical Sciences*. 43(4), October–December 1995.

NEAR instruments and science goals. *Journal of Geophysical Research (Planets)*. 102(E10), October 25, 1997.

NEAR's initial findings from Mathilde flyby. *Science*. 278(5346), December 19, 1997.

NEAR's detailed results from Mathilde. *Icarus*. 140(1), July 1999.

NEAR's findings from the 1998 Eros flyby. *Science*. 285(5427), July 23, 1999.

NEAR's initial findings from Eros orbit. *Science*. 289(5487), September 22, 2000.

NEAR's closest orbital encounter with Eros. *Science*. 292(5516), April 20, 2001.

NEAR's landing and highest resolution imaging. *Nature*. 413 (6854), September 27, 2001.

NEAR's detailed results from Eros orbit. *Icarus*. 155(1), January 2002, and *Meteoritics and Planetary Science*. 36 (12), December 2001.

OTHER TECHNICAL PAPERS

Report of the Near Earth Asteroid Rendezvous (NEAR) Science Working Group. JPL Report 86-7. Jet Propulsion Laboratory, Pasadena, California, June 1986.

Dunham, D. W. *et. al.* Recovery of NEAR's Mission to Eros. *Acta Astronautica*. 47(2–9), 2000, 503–512.

Izenberg, N. R. and Anderson, B. J. NEAR swings by Earth en route to Eros. *EOS Transactions*. 79(25), June 23, 1998, 289, 294–295.

Scheeres, D. J. *et. al.* Mission design and navigation of NEAR's encounter with asteroid 253 Mathilde. *Advances in the Astronautical Sciences*. 99, 1998, 1157–1173.

WEB SITES

Official NEAR home page. http://near.jhuapl.edu

NASA's Near Earth Object Study Program page. http://neo.jpl.nasa.gov/

IAU Minor Planet Center page. http://cfa-www.harvard.edu/cfa/ps/mpc.html

Asteroid and Comet Impact Hazards page. http://impact.arc.nasa.gov/

Glossary of technical terms and abbreviations

Words in *italic* have their own entries elsewhere in the glossary.

albedo The proportion of incident light that is reflected from the surface of a body.

anticoincidence shield A part of the gamma-ray spectrometer that helps to focus radiation entering the instrument so that only gamma rays that enter the detector at the same time from anywhere within the field of view are detected. All other incident radiation is rejected from the instrument.

aphelion For an object orbiting the Sun, the time when or place where it is farthest from the Sun.

APL Abbreviation for the Applied Physics Laboratory of Johns Hopkins University, Baltimore.

array An arrangement of individual radiation detectors in the form of a square, rectangle, or other geometrical shape.

asteroid belt The region in the solar system between the orbits of Mars and Jupiter at a distance of 2.0–3.3 *AU* from the Sun where the orbits of the vast majority of asteroids are located.

astronomical unit A unit of measurement used in astronomy for distances in the solar system. It is the average distance between Earth and the Sun: 149 597 870 km (92 955 730 miles).

AU Abbreviation for *astronomical unit*.

Cambrian epoch Between about 590 and 505 million years ago; the earliest geological period of the Paleozoic era.

carbonaceous chondrites Rare stony meteorites with a composition very similar to the Sun's (apart from the most volatile elements). They are thought to consist of some of the most primitive material in the solar system, unaltered since the time when the planets were forming.

CCD Abbreviation for charge-coupled device, a sensitive electronic imaging device widely used in astronomy. A CCD detector consists of an array of picture elements (pixels) made of semi-conducting silicon.

chondrite A common type of stony meteorite, nearly all of which contain chondrules – small spheres (less than 1 mm to several mm in size) composed of primitive silicate minerals or glass.

cosmic radiation High-energy elementary particles that travel through the universe at almost the speed of light.

cosmochemistry The study of the chemistry and composition of astronomical bodies and of the universe as a whole.

C-type asteroids A category of dark, grayish, asteroids with *albedos* of around 5%. They are the most common type in the outer part of the *asteroid belt*. C stands for "carbonaceous".

Deep Space Network (DSN) Three 70-m (230-feet) radio antennas used by NASA to communicate with spacecraft. They are located at Goldstone in California, near Madrid in Spain and near Canberra in Australia. See page 88 in Chapter 8.

Delta-V (ΔV) The change in a spacecraft's velocity achieved at launch or as the result of an engine burn during a maneuver; the potential for such a velocity change contained in a supply of fuel on board a spacecraft.

Doppler shift The change between the emitted and received wavelength of radiation as a result of relative motion between source and observer in the direction of the line joining them. If source and observer are moving towards each other, the shift is to shorter wavelengths (blue shift). If they are moving apart, the shift is to longer wavelengths (red shift).

DSN Abbreviation for *Deep Space Network*.

DSWG Abbreviation for Discovery Science Working Group.

ejecta Material excavated by an impact (or thrown out by a volcano), which either gets deposited elsewhere on the surface of the body or escapes into space.

FBC Abbreviation for "Faster, Better, Cheaper", a philosophy for planning NASA space missions introduced in the early 1990s.

fluxgate magnetometer A type of instrument for detecting weak magnetic fields, based on measuring small changes in the current of an electrical circuit operating within the magnetic field to be measured. NEAR's magnetometer uses a fluxgate circuit.

gamma-ray bursts Powerful bursts of gamma rays and X-rays from cosmic sources, lasting between a few milliseconds and a few tens of seconds. Their origin is not known for sure, but they come from remote galaxies and briefly emit more energy than a million galaxies combined. On average, two are detected every day.

gas proportional counter The type of X-ray detector used in NEAR's X-ray/gamma-ray spectrometer instrument. X-rays emitted from the surface of Eros enter a small chamber filled with argon and methane gas and induce a small current that is proportional to the energy of the incident X-rays. This in turn can be used to infer the elemental composition of the surface.

gravity assist The use of the gravitational field of a planet to change the speed and direction of a spacecraft.

GRS Abbreviation for the NEAR mission's gamma-ray spectrometer.

HED meteorite A member of one of three related classes of meteorites – howardites, eucrites, and diogenites – which, on the basis of spectroscopic evidence, appear to come from the asteroid 4 Vesta.

IAU Abbreviation for the International Astronomical Union, a non-governmental international body with headquarters in Paris, which was established in 1919 to foster international cooperation in astronomy.

JPL Abbreviation for the Jet Propulsion Laboratory, an institution in Pasadena, California, operated by the California Institute of Technology in support of NASA programs.

keV Abbreviation for kilo-electron-volt, a unit of energy used particularly in connection with elementary particles and X-rays. X-rays are frequently characterized by their energy rather than their wavelength. 1 keV corresponds to a wavelength of 0.124 nanometers.

Kuiper belt (also called the Edgeworth–Kuiper belt) A belt of small bodies, similar to asteroids but containing a significant proportion of ices, occupying a region in the solar system extending from the orbit of Neptune (30 AU from the Sun) out to about 150 AU from the Sun.

LAF Abbreviation for low-altitude flyover.

laser rangefinder An instrument that uses the reflection of laser light to determine its distance from a surface.

lidar A lasar system for measuring distances or altitudes; derived from LIght Detection And Ranging by analogy with radar (RAdio Detection And Ranging).

MAG Abbreviation for the NEAR mission's magnetometer.

main–belt asteroid An asteroid whose orbit is currently within the *asteroid belt*.

MD Abbreviation for the NEAR mission's design team.

meteoroid A small body in space, especially one that might enter Earth's atmosphere to be observable as a meteor or land as a meteorite. The term meteoroid is generally used for objects of the order of 10 meters across or smaller.

MeV Abbreviation for mega-electron-volt, a unit of energy used particularly in connection with elementary particles and gamma rays. Gamma rays are frequently characterized by their energy rather than their wavelength. 1 MeV corresponds to a wavelength of 1.24×10^{-4} nanometers.

micron Alternative word for micrometer, a unit of measurement equal to one thousandth of a millimeter.

moment of inertia For a body rotating about an axis, a measure of the way mass is distributed within the body in relation to the rotation axis.

MOPS Abbreviation for the NEAR mission operations team.

MSI Abbreviation for the NEAR mission's multispectral imager, or the team operating it.

NaI scintillation detector An instrument that converts the flashes of light emitted by sodium iodide (NaI) when radiation falls upon it into an electrical signal that indicates the intensity and energy of the radiation.

NAV Abbreviation for the NEAR mission's navigation team.

NEA Abbreviation for *near Earth asteroid*.

near Earth asteroids (NEAs) Asteroids able to approach exceptionally close to Earth in their present orbits. Most are in orbits with *perihelion* distances less than 1.3 AU and *aphelion* distances greater than 0.983 AU. It is believed that they originally belonged to the *asteroid belt* but were perturbed into new orbits by the gravitational attraction of one or more planets.

NIS Abbreviation for the NEAR mission's near-infrared spectrometer.

NLR Abbreviation for the NEAR mission's laser rangefinder.

nm Abbreviation for nanometer, a unit of measurement equal to one billionth (10^{-9}) of a meter.

OD Abbreviation for the NEAR mission's orbit determination team.

olivine Magnesium iron silicate, a mineral commonly found in meteorites.

Oort (comet) cloud A hypothesized spherical shell of billions to trillions of small icy bodies surrounding the solar system at a distance of about 1 light year (50 000 AU). It has never been observed directly but, with the *Kuiper belt*, is thought to be the "reservoir" from which comets seen in the inner solar system originate.

opnav Abbreviation for optical navigation (image).

ordinary chondrite A type of meteorite: the most common of three classes of *chondrites*.

parallax The apparent change in position of an object on the sky (measured as an angle) when there is a change in the relative position of the object and its observer. Astronomers sometimes use "parallax" as a synonym for "distance" since the measurement of parallax is a fundamental way of determining the distance of an astronomical object.

perihelion For an object orbiting the Sun, the time when or the place where it is closest to the Sun.

phase angle The angle between the line connecting a body to the Sun and the direction from which it is being observed.

plagioclase A class of silicate minerals that are major constituents of meteorites.

polarimetry Measuring the state of polarization of light or other electromagnetic radiation. In polarized radiation, the electric fields associated with the photons are not oriented at random. Reflection from different kinds of surfaces affects the polarization of light in different ways, so polarimetry is a means of investigating the nature of a surface.

pond On Eros, an approximately circular and flat deposit of relatively fine material.

PRF Abbreviation for positive relief feature, an otherwise unidentified protuberance on the surface of Eros.

pyroxene Magnesium iron calcium silicate, a mineral commonly found in meteorites.

radial velocity The velocity of an object relative to an observer along the line of sight, either towards or away from the observer.

radiometry Measuring the intensity of light, or other electromagnetic radiation, emitted or reflected by an object.

regolith A surface layer of loose rocky debris formed as a result of impacts.

resonance fluorescence The basic physics principle behind the operation and interpretation of data from the NEAR XGRS instrument. High-energy solar wind or cosmic ray particles striking the surface of Eros induce the emission (fluorescence) of X-rays and gamma rays from the surface in a way that is diagnostic of the specific elements that make up the surface.

solar nebula The cloud of interstellar gas and dust that condensed to form the solar system about five billion years ago.

solar wind The stream of elementary particles, chiefly protons and electrons, flowing out from the Sun into interplanetary space at speeds of up to several hundred kilometers per second.

space weathering Alteration of the properties of minerals on the surface of an airless body by the action of meteorites of microscopic size, cosmic radiation, ultraviolet radiation, etc.

spectrometer An instrument for dispersing electromagnetic radiation into a spectrum and measuring features in the spectrum. The spectrum of radiation emitted from or reflected by an object may be used to deduce the object's composition or mineralogy.

S-type asteroids A diverse class of asteroids with intermediate *albedos* (10–30%), which are common in the inner part of the *asteroid belt*. S stands for silicate, since, from spectroscopic evidence, they are thought to contain substantial amounts of silicate minerals.

tectonic Caused by the movement of the rocky fabric of a body, such as folding or faulting.

three-axis stabilized A system for controlling a spacecraft by which the yaw, pitch, and roll are precisely controlled during the flight using either rocket thrusters or other methods. Controlling the spacecraft's orientation and preventing it from spinning or tumbling are required to properly point cameras and other instruments at their targets, as well as to navigate the spacecraft through interplanetary space.

trans-Neptunian object An object in the *Kuiper belt*.

undifferentiated Describing a body that has not been stratified into layers of different composition but is the same throughout.

V-type asteroids Asteroids with spectra like those of 4 Vesta. Apart from Vesta itself, the only others known are small objects (5–7 km across) in the vicinity of Vesta, which are probably fragments of Vesta ejected by a massive impact.

X-band telemetry system The method used by NEAR Shoemaker to transmit and receive commands and data to and from the NASA *Deep Space Network*. X-band radio signals have a frequency around 10 000 MHz (wavelength around 3 cm).

XGRS Abbreviation for the X-ray and gamma-ray spectrometer instrument package on the NEAR Shoemaker spacecraft, or the team operating it.

XRS Abbreviation for NEAR's X-ray spectrometer.

Index